Second Edition

Field Hydrogeology

A Guide for

Site Investigations

and

Report Preparation

Second Edition

Field
Hydrogeology

A Guide for

Site Investigations

and

Report Preparation

John E. Moore

CRC Press
Taylor & Francis Group
Boca Raton   London   New York

CRC Press is an imprint of the
Taylor & Francis Group, an **informa** business

CRC Press
Taylor & Francis Group
6000 Broken Sound Parkway NW, Suite 300
Boca Raton, FL 33487-2742

First issued in paperback 2017

© 2012 by Taylor & Francis Group, LLC
CRC Press is an imprint of Taylor & Francis Group, an Informa business

No claim to original U.S. Government works
Version Date: 20110613

ISBN 13: 978-1-138-07714-0 (pbk)
ISBN 13: 978-1-4398-4124-2 (hbk)

Visit the Taylor & Francis Web site at
http://www.taylorandfrancis.com

and the CRC Press Web site at
http://www.crcpress.com

Contents

Preface

This document was prepared as a practical instruction guide for field hydrogeologists. It differs from other field guides in that the American Society for Testing and Materials (ASTM) and U.S. Geological Survey (USGS) techniques manuals and environmental protection manuals were used to prepare this guide and are referenced extensively. The guide was written in part to expose hydrogeologists, particularly those new to field investigations, to these detailed and useful standard operating procedures (SOPs).

This second edition of *Field Hydrogeology* contains the following new sections:

- history of hydrogeology
- field safety
- groundwater quality and testing the quality of groundwater
- construction of hydrogeologic cross sections and maps
- federal laws to protect groundwater
- transboundary aquifers
- new THEIS computer model for designing aquifer tests

This edition also has new case studies: Nubian aquifer in northern Africa, Mexico City, and the High Plains aquifer in the southwestern United States. Another addition to this field guide is a computer program to design the location of observation wells for aquifer tests.

I became interested in SOPs when I was an electronics officer in the U.S. Air Force and later a teaching assistant at the University of Illinois. As an electronics officer I wrote the SOP for using electronic testing instruments. At the university, I developed standard procedures for sedimentary petrology laboratory analyses. As a hydrologist with the USGS and as senior hydrogeologist for an environmental consulting firm, I prepared many standard hydrogeologic field procedures. I believe that these procedures are an essential part of an investigation because they include the application of consistent methods of data collection and they provide guidance to beginning hydrogeologists and reminders to experienced hydrogeologists. I visited an environmental consulting firm recently and asked their senior hydrogeologist what he thought about ASTM guides and standards. He felt that some were too stringent and some inhibited creative thinking, but he admitted that overall they have improved the quality and consistency of field data.

The USGS took the lead in establishing standards for collecting hydrologic data when they published the USGS Techniques of Water Resources Investigation series and their books on recommended methods for collecting hydrologic data.

ASTM has been in operation for 103 years. It has grown into one of the largest voluntary standards systems in the world and currently has more than 33,000 members. More than 9,500 standards are published by ASTM annually. It provides a forum for producers, users, consumers, and those with a similar interest, such as representatives of government and academia, to have a common ground to developed standards for materials, products, and services. ASTM standards are incorporated into contracts; scientists and engineers use them in laboratories; designers use them for specifications and plans; and government agencies reference them in guidelines, regulations, and laws. ASTM standards are developed and utilized voluntarily. They become legally binding only when a government body references them in regulations or they are cited in a contract. The major ASTM committees dealing with water problems are Soil and Rock, Water, Waste Management, Biological Effects and Environmental Fate, and Assessment. There are more than 100 hydrogeology standards. It is the obligation of every hydrogeologist to evaluate each standard for validity and utility before using it. Additional downloadable material can be found on the book's website at www.crcpress.com/product/isbn/9781439841242.

CSUPAWE

The computer model program written by Dr. Daniel K. Sunada (Colorado State University) has a designation of CSUPAWE. The program can be used to calculate the water level response to withdrawal of groundwater and artificial recharge.

THEIS COMPUTER MODEL PROGRAM

The computer model prepared by John McCain (USGS) is called THEIS, and is simpler than CSUPAWE. The THEIS program solution calculates drawdown due to a pumping well at a constant rate. The input and output can be in metric or English (U.S.) units. The program is distributed as freeware.

Acknowledgments

I would like to thank the hydrologists who provided guidance to me in the preparation of this book. I am grateful to Joe Rosenshein, Joe Downey, and Edwin D. Gutentag, who gave me the course material they used for the courses they taught for ASTM and gave me encouragement to write this guide. Many hydrologists generously contributed expertise and time in the development of this book. My special thanks go to Bozidar Biondic, Charles Morgan, Jim LaMoreaux, Mario Cuesta, Phil E. LaMoreaux, and A. Ivan Johnson. Chester Zenone, Patrick Tucci, and James R. Lundy who provided outstanding editorial and technical reviews of the book. I also appreciate the support of the International Association of Hydrogeologists Commission on Education and Training, Taylor & Francis Publishing Group, my students at Metro State College, and attendees at field hydrogeology short courses in Denver, Albuquerque, St. Paul, Lisbon, Colombia, and Munich. Jack Sharp, Van Brahana, Mike Wireman, and J. Joel Carrillo-Rivera provided chapter material for this book. I could not have completed this work without the support of my wife, Dr. Unetta T. Moore, my son Sean, and my daughter Pamela.

John Moore

Acknowledgments

About the Authors

John E. Moore, Ph.D., is an internationally recognized research scientist and hydrogeologist. He has earned a B.A. in geology from Ohio Wesleyan University and a Ph.D. from the University of Illinois. He is currently an adjunct professor at Metro State College in Denver, Colorado, and presents short courses for the Geological Society of America and the International Association of Hydrogeologists. He has more than 50 years of experience as scientist, technical advisor, and senior hydrologist with the Water Resources Division of the U.S. Geological Survey (USGS) and the Environmental Protection Agency (EPA). At the USGS he planned and directed hydrologic investigations, supervised well drilling, designed aquifer tests, investigated contamination, supervised writing and review of 1,500 reports each year, and presented hydrologic short courses in project planning, report review, and field hydrogeology. Moore has served as an advisor to the EPA, Department of Energy, Department of Defense, U.S. Congress, and the State of Colorado. He is past president of the American Institute of Hydrology (AIH) and the International Association of Hydrogeologists (IAH), and associate editor of *Environmental Geology.*

Dr. Moore received the Department of Interior Meritorious Service Award, AIH Founders award, IAH Honorary Members Award, and the National Groundwater Association Life Member Award. He is the author of 7 books and 50 scientific articles. Because of his contributions to hydrology and publications, he was presented an honorary doctor of science on October 5, 2010, at Ohio Wesleyan University.

J. Joel Carrillo-Rivera, Ph.D., earned an M.Sc. in hydrogeology from University Colleague in Great Britain and a Ph.D. in geology (hydrogeology) from London University. He is a researcher at the Institute of Geography, UNAM; Lower House Advisor; member of the National Academy of Sciences; researcher and reviewer of CONACyT; European Community external advisor; and past president of the Mexican chapter of the International Association of Hydrogeologists. Teaching hydrogeology and tutoring thesis work in groundwater have been part of his major responsibilities in the Institute of Geography at the University of Mexico. The publication of books, maps, and papers has been his important academic activity.

Michael Wireman, M.S., is a hydrogeologist employed by the U.S. EPA in Denver, Colorado, where he serves as a national groundwater expert. In his current position he provides technical and scientific support to several EPA programs, other federal agencies, international programs, and groundwater protection/management programs in several Western states. He has significant experience in the legal, scientific, and programmatic aspects of groundwater resource management. He also has extensive experience in groundwater work in the Baltic countries, Ukraine, Romania, and Georgia. He has served as an adjunct professor at Metropolitan State College in Denver, where he taught contaminant hydrology, and he teaches a class on  fractured rock hydrology for the National Groundwater Association. He is a member of the Colorado Groundwater Association, the National Groundwater Association, and the Geological Society of America; and the chairman of the American Chapter of the International Association of Hydrogeologists.

Key Hydrogeology Titles

SOME KEY ASTM HYDROGEOLOGY STANDARDS AND GUIDES

ASTM, 1996, *ASTM design and planning for ground water and vadose zone investigations*, 126 pp.

ASTM, 1996, *ASTM standards on analysis of hydrologic parameters and ground water modeling*, 144 pp.

ASTM, 1996, *ASTM standards on ground water and vadose zone investigations: drilling, sampling, well installation and abandonment procedures*, 247 pp.

ASTM, 1997, *ASTM standards relating to environmental site characterization*, 1410 pp.

ASTM, 1999, *ASTM standards on design, planning and reporting of ground water and vadose zone investigations*, second edition, 600 pp.

ASTM, 1999, *ASTM standards on determining subsurface hydraulic properties and ground water modeling*, second edition, 320 pp.

SOME KEY USGS PUBLICATIONS IN HYDROGEOLOGY

Alley, W. M., Riley, T. E., and Franke, O. L., 1999, *Sustainability of ground-water resources*, U.S. Geological Survey Circular 1186, 79 pp.

Buchanan, T. J., and Somers, W. P., 1969, Discharge measurements at gaging stations, in *USGS techniques of water-resources investigations*, TWRI Book 3, Chapter A8, 65 pp.

Heath, R. C., 1984, *Basic ground-water hydrology*, U.S. Geological Survey Water-Supply Paper 2220, 84 pp.

Hem, J. D., 1985, *Study and interpretation of the chemical characteristics of natural water*, U.S. Geological Survey Water-Supply Paper 2254, 263 pp.

Keys, W. S., 1989, Borehole geophysics applied to ground water investigations, in *USGS techniques of water-resource investigations*, TWRI Book 2, Chapter E2.

Lohman, S. W., 1972, *Ground water hydraulics,* U.S. Geological Survey professional paper, 708 pp.

Reilly, T. C., and others, 2008, *Ground-water availability in the United States*, U.S. Geological Survey Circular 1323, 70 pp.

Stallman, R. W., 1971, Aquifer-test, design, observations, and data analysis, in *USGS techniques of water-resources investigations*, TWRI Book 3, Chapter B1, 26 pp.

Taylor, C. J., and Alley, W. H., 2001, *Ground-water level monitoring and the importance of long-term water level data*, U.S. Geological Survey Circular 1217, 68 pp.

U.S. Geological Survey, 1998, *Ground water atlas of the United States: Introduction and national summary.*

U.S. Geological Survey, 2002, *Concepts for national assessments of water availability and use*, U.S. Geological Survey Circular 1223, 343 pp.

Winter, T. C., and others, 1998, *Ground water and surface water: A single resource*, U.S. Geological Survey Circular 1139, 76 pp.

Wood, W. W., 1976, Guidelines for collection and field analysis of groundwater samples for selected unstable constituents, in *USGS techniques of water-resources investigations*, TWRI Book 1, Chapter D2, 53 pp.

Zody, A. A. R., Eaton, G., and Mabey, D. R., 1974, Application of surface geophysics to ground-water investigations, in *USGS techniques of water-resource investigations*, TWRI Book 2, Chapter D1, 116 pp.

SOME KEY U.S. EPA PUBLICATIONS IN HYDROGEOLOGY

EPA, 1985, *Practical guide for ground-water sampling*, EPA/600/2-85/104.

EPA, 1991, *Site characterization for subsurface remediation*, EPA/625/4-91/026.

EPA, 1993, *Subsurface characterization and monitoring techniques: a desk reference guide, Volume I, Solids and ground water appendices: A and B.*

EPA, 1998, *The national water quality inventory report to Congress.*

1 Introduction

Hydrogeologic field methods are an essential element to the science of hydrology and essential in the training of hydrogeologic students. Groundwater field and theoretical studies rely on data. It is important that students appreciate how data are collected, the uncertainties in data collection, and data interpretation. They need to understand the critical data needs of the project. The field experience can amplify the theoretical concepts taught in the classroom.

This book presents current standard methods and guides for planning and undertaking field investigations. It covers concepts, aquifer identification, groundwater movement, recharge, discharge, and rules for professional conduct. Information is presented on hydrogeologic principles, conceptual models, sources of hydrologic information, geophysical methods, surface investigations, subsurface investigations, well inventory, design of aquifer tests, stream flow measurements, report planning, report writing, and report review.

Key features of the book arc as follows:

- Uses ASTM standards and guides for fieldwork
- Includes a computer program to design aquifer tests
- Describes U.S. Geological Survey (USGS) and U.S. Environmental Protection Agency (EPA) field techniques
- Provides guides for plan site investigation plans and site reports
- Contains detailed list of source of information and websites.
- Describes basic groundwater principles and concepts (groundwater introduction)
- Contains checklists for the preparation and review of hydrogeologic reports

The book was the basis for short courses presented at International Association of Hydrogeologists (IAH) meetings in Germany, Portugal, Colombia, China, Argentina, and Slovenia (IAH Congresses). Short courses also were presented at meetings of the Geological Society of America, National Groundwater Association, and American Institute of Hydrology.

This book was prepared as a practical guide. It will help field personnel to evaluate commonly encountered problems in hydrogeologic investigations. The emphasis in this field guide is placed on a practical how-to rather than a textbook approach. Hydrogeologic investigations have changed in the past 15 years from resource evaluations to aquifer sustainability. Therefore, this guide will place emphasis on that aspect of field studies. Major source references for this guide were USGS *Techniques of Water-Resources Investigations* and ASTM standards and guides for soil, rock, and groundwater. A basic review of hydrogeologic concepts follows.

1.1 HYDROGEOLOGIC CONCEPTS

Groundwater is one of the nation's most important natural resources. It is the principal source of drinking water for about 50 percent of the U.S. population, providing approximately 96 percent of water used for domestic supplies and 40 percent of the water used for public supplies (Hutson and others 2004). It is a significant source of the nation's water resources. Much of the flow in streams and the water in lakes and wetlands is sustained by groundwater discharge.

Groundwater is difficult to visualize. Some people believe that groundwater collects in underground lakes or rivers. In fact, groundwater is simply the subsurface water that fully saturates pores or cracks in soils and rocks. Such conditions may exist in cavernous limestone or lava rock, but these are relatively uncommon. Most groundwater is contained in and moves through the pore spaces between rock particles or in fractures and fissures in rocks. When the pore spaces in sand and gravel become saturated with water, the water is groundwater.

Groundwater is replenished by precipitation and, depending on the local climate and geology, is unevenly distributed (Figure 1.1). When it rains some of the water runs off to streams, some evaporates, and some recharges aquifers (Moore and others 1995).

Groundwater occurs nearly everywhere in the world at depths ranging from land surface to about 1,500 m below the land surface. Most groundwater is present within 1,500 m of the land surface. Below about 1,500 m most pores and cracks are closed because of the weight of overlying rocks. Most groundwater is obtained within 400 m of the land surface.

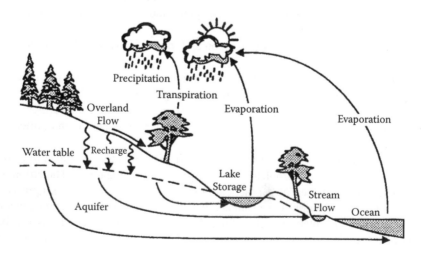

FIGURE 1.1 Hydrologic cycle. (From Moore, J., *Field Hydrology: A Guide for Site Investigations and Report Preparation*, CRC Press, Boca Raton, 2002. With permission from Taylor & Francis.)

After the water requirements for plants and soil are satisfied, any excess water will infiltrate to the water table. Below the water and toward to streams, springs, or wells.

Groundwater is stored and transmitted by aquifers. One simple definition of an aquifer is a formation, or part of a formation, that contains sufficient saturated permeable material to yield usable quantities of water to wells and springs. The word *aquifer* comes from two Latin words: *aqua*, or water, and *ferre*, to bear or carry. Aquifers literally carry water underground. An aquifer may be a layer of gravel or sand, sandstone, limestone, lava flow, or fractured granite.

The major productive aquifers in the world are unconsolidated sand and gravel, limestone, dolomite, basalt, and sandstone. The location and yield of aquifers are dependent on geologic conditions, such as the size and sorting of grains in unconsolidated deposits, faulting, solution openings, and fracturing in consolidated rocks (sandstone, basalt, limestone, fractured granite, volcanic ash).

The properties of aquifers that affect the storage and flow of groundwater are their thickness, transmissibility, porosity, storage capability, and hydraulic conductivity. Methods for determining these properties are described in Freeze and Cherry (1979).

The quantity of water that a given type of soil, sediment, or rock can hold depends on the porosity of the formation. Porosity is a measure of pore space between the grains of the rock or of cracks in the rock that can fill with water. All rocks and soils contain pore spaces. The percentage of the total volume of the soil or rock that consists of pores is its porosity. If the porosity of sand is 30 percent, then 30 percent of the total volume of a quantity of sand is pore space and 70 percent is solid material. The porosity of rocks varies widely. In granite, the porosity is typically less than 1 percent. In unconsolidated sand and gravel, it may be as great as 30 percent.

Porosity is either primary or secondary (Moore and others 1995). Primary porosity, such as pores between sand grains, is created when rocks are formed (Figure 1.2). Secondary porosity is created by the solution of rocks such as limestone and by fracturing of rocks.

Porosity defines the storage capacity of an aquifer. The shape, sorting, and packing of grains control primary porosity. Sediment is poorly sorted when the

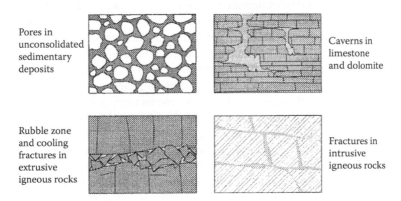

Pores in unconsolidated sedimentary deposits

Caverns in limestone and dolomite

Rubble zone and cooling fractures in extrusive igneous rocks

Fractures in intrusive igneous rocks

FIGURE 1.2 Aquifer pore spaces. (From Moore, J., *Field Hydrology: A Guide for Site Investigations and Report Preparation*, CRC Press, Boca Raton, 2002. With permission from Taylor & Francis.)

grains are not all the same size and thus the spaces between the larger grains are filled with smaller grains. When the pores are not connected and dead-end pores exist, there is no groundwater flow. This condition would be analogous to water in a sponge. On the other hand, effective porosity refers to pores that are interconnected. Cementation of sand may isolate pores and thereby reduce its effective porosity. Clays have many pores but do not yield water readily because the pores are not effectively interconnected. Secondary porosity, which is present in joints, fractures, solution openings, and openings created by plants and animals, develops after rocks are formed. The number and arrangement of fracture openings and the degree to which they are filled by finer grained material determine secondary porosity.

If water is to move through rock, the pores must be connected to one another. If the pore spaces are connected and large enough that water can move freely through them, the rock is said to be permeable. Permeability and hydraulic conductivity are measures of an aquifer's ability to transmit water. An aquifer with large hydraulic conductivity is highly permeable, can easily transmit water, and will yield a large volume of water to wells or springs.

Hydraulic conductivity and hydraulic gradient determine the rate of flow of groundwater. The hydraulic gradient is the change in head (decrease in water level) per unit distance. The greater the hydraulic conductivity, the less the resistance to flow. In volcanic or crystalline rocks and in some carbonate rocks, the hydraulic conductivity depends on the size of the openings in the rock and how well the cracks or fractures are interconnected. In sand and gravel aquifers, hydraulic conductivity depends partly on grain size; coarse-grained materials, such as coarse sand and gravel, have greater hydraulic conductivity than fine-grained materials, such as fine sand or clay. Therefore, rocks and unconsolidated sediments with high hydraulic conductivity make the most aerially productive aquifers.

Porosity and hydraulic conductivity are heterogeneous. An aquifer that consists of similar rock or is made up of similar size, and has the same properties at all locations would be considered homogeneous.

Geologic depositional (sedimentation) patterns and materials can affect groundwater movement. Alluvial aquifers commonly show a pronounced difference, or anisotropy, between horizontal and vertical hydraulic conductivity. Horizontal hydraulic conductivity is usually much greater than vertical hydraulic conductivity (as much as 100 to 1) due to the vertical alternation of sand and clay lenses (stratification). Fractured-rock hydraulic conductivity is anisotropic because in both the vertical and horizontal directions the hydraulic conductivity is completely governed by the orientations of fractures. An aquifer that has the same hydraulic properties in every direction is isotropic (homogeneous); however, this type of aquifer is rare.

Aquifers are classified as unconfined and confined. The water table is the upper boundary of an unconfined aquifer. Recharge to unconfined aquifers takes place primarily by downward seepage through the unsaturated zone. The water table in an unconfined aquifer rises or declines in response to recharge from rainfall and declines during periods of no recharge as the groundwater moves downgradient toward stream valleys (or is pumped through wells). When a well tapping an unconfined aquifer is pumped, the water level is lowered, gravity causes water to flow to the well, and sediments near the well are dewatered. Unconfined aquifers are usually the uppermost aquifers and, therefore, are more susceptible to contamination from activities occurring at the land surface.

A confined, or artesian, aquifer contains water under pressure greater than atmospheric conditions when sediments of lower permeability (such as clay or shale) overlie the aquifer. The overlying low-permeability layer is called a confining bed. A confining bed has very low permeability and is poorly transmissive to groundwater. It thus restricts the movement of groundwater either into or out of the aquifer. Due to the pressure on the water in a confined aquifer, water levels in wells that tap a confined aquifer will rise above the top of the aquifer. If the water level in a well tapping a confined aquifer is above the land surface, the well is called a flowing artesian well.

After entering the aquifer, water moves downgradient and eventually is discharged from the aquifer by springs, seeps, lakes, or streams.

1.2 AQUIFERS

Sand and gravel aquifers produce most of the groundwater pumped in many parts of the world including North America, the Netherlands, France, Spain, and China. Sand and gravel aquifers are common near large to moderately sized streams; these aquifers were formed by the deposition of sediment by rivers or the meltwater streams from glaciers. Other sand and gravel aquifers are the result of erosion and subsequent redeposition of sediments from mountain ranges into adjacent basins, and comprise many of the major aquifers in the western United States.

Sand grains with silica or calcium carbonate cement form sandstone aquifers. Their porosity ranges from 5 to 30 percent. Their hydraulic conductivity is a function of grain packing, grain size, and amount of cement (clay, calcite, and quartz). Sandstone is an important source of groundwater in Libya, Egypt (Nubian sandstone), Great Britain (the Permian-Triassic sandstones), the north central United States (St. Peter–Mount Simon sandstone), and in the west central United States (the Dakota sandstone).

Limestone aquifers are the sources of some of the largest well and spring yields. Openings that existed at the time the rocks were formed are commonly enlarged by solution (dissolved by water), providing highly permeable flow paths for groundwater. Limestone is an important source of water in many parts of the world such as the Texas–Edwards aquifer, Montana–Madison aquifer, and Illinois–Galena aquifer.

Basalt and other volcanic rocks also make up some of the most productive aquifers. Basalt aquifers contain water-bearing spaces in the form of shrinkage cracks, joints, interflow zones, and lava tubes. Lava tubes are formed when tunneling lava ceases to flow and drain out, leaving a long, cavernous formation. Volcanic rocks form important aquifers in Hawaii, Nevada, Idaho, Mexico, Central America, and India.

Fractured igneous and metamorphic rocks are the principal sources of groundwater for people living in many mountainous areas. Where fractures are numerous and interconnected, these rocks can supply water to wells and can be classified as aquifers. Wells in these rocks are commonly less than 30 m deep. In igneous and metamorphic rocks, much of the hydraulic conductivity occurs in the deeply weathered zone and in fractures enlarged by weathering; examples of such aquifers are in Piedmont plateau at the eastern border of the Appalachian Mountains in the United States.

1.3 GROUNDWATER MOVEMENT

The two principal functions of aquifers are to store and to transmit water. Groundwater is transmitted through aquifers along the hydraulic gradient from areas of high head to areas of low head due to the driving force of gravity, generally (for unconfined aquifers) conforming to the slope of the land surface. It is very important that planners and well owners understand this concept.

Geologic conditions (type and sequence of rocks) in the subsurface can have a major control on the direction and rate of groundwater flow. Groundwater movement is very slow when compared with that of surface water. Surface-water flow is usually measured in meters per second and groundwater flow in centimeters per year.

Most of the water in an aquifer infiltrates the ground within a radius of a few tens of kilometers from where it is tapped by wells, consumed by vegetation, or discharged to the land surface. Except for the water in some very deep aquifers, groundwater does not travel for hundreds of miles. After entering the aquifer, the water moves downgradient, under gravity or pressure to discharge into seeps, springs, streams, wells, saline soils, wetlands, lakes, or the sea.

As noted above, groundwater flows in response to an energy gradient; that is, it moves or flows from areas of high energy or head to low energy. Head is composed of two parts, the pressure head that produces the column of water above the open interval in the well, and the elevation head. The elevation of the top of the open interval is measured relative to a datum (usually mean sea level). Depth to water is normally measured from a reference point (top of casing) that has been surveyed to establish a precise elevation. The density effect of the water implies that the denser a water column the higher the head to be observed at the bottom of the system (Carrillo-Rivera and Cardona 2008). This is a crucial factor to consider if groundwater measurements are made close to the sea, where denser sea water is to influence head computations of (fresh) water-table levels. Similar density considerations are to be given to water-table measurements (with cold water of ≈20°C) influenced by higher temperature groundwater (water at depth with 70°C), which occurs when local and regional flows are tapped by a well constructed on a thick lithological sequence. Note that the well is not necessarily required to reach the bottom of the system (>1,500 m) to induce thermal water. These data are used to compute water-level elevation. Although the depth to water is useful to know, without converting water depths from several wells to hydraulic head, the direction and rate of groundwater movement cannot be determined without converting water depths from several wells to hydraulic head.

The first systematic study to define the movement of water through porous materials was performed in 1856 by Henry Darcy, a French engineer. He observed that the volumetric rate of water flow (Q) through a bed of sand is directly proportional to the amount of force on the water across area (A), that is, the difference in pressure head (h), and is inversely proportional to the length of the flow path (l). The quantity of flow is proportional to a coefficient K (hydraulic conductivity). Darcy's Law is sometimes written Q = KA h/l, where A is the cross-sectional area through which the groundwater flows.

Heads measured in wells tapping an unconfined aquifer are used to construct a water-table contour map, which is analogous to a topographic map but represents the slopes of a water surface rather than of a land surface. Water flows from high head to low head (high elevation to lower elevation). This map helps the hydrogeologist to define groundwater recharge, movement, and discharge. Groundwater typically follows a path that is approximately perpendicular to water-table contours and is referred to as a flow line.

Head, measured in wells tapping a confined aquifer, is used to construct a water-table contour or potentiometric-surface map. It is analogous to a topographic map but represents the slopes of a pressure water surface rather than of a land surface. Water flows from high head to low head (high elevation to lower elevation).

The rate at which groundwater moves through an aquifer ranges from only a fraction of a few millimeters per day (sand and gravel) to several meters per day (cavernous limestone). In a fractured permeable rock with small effective porosity, velocity can be relatively large. In contrast, it can take decades or centuries for groundwater to pass through low-permeable aquifer beds within the same area.

1.4 RECHARGE AND DISCHARGE

Groundwater in unconfined aquifers moves from topographically high areas (recharge) to topographically low areas (discharge). In a recharge area, the potential energy decreases with depth, which results in downward movement of water. Between the recharge and discharge areas, groundwater flow is primarily horizontal but with some small slope. In a discharge area, the potential energy increases with depth, resulting in upward movement of groundwater.

In humid areas with porous soils, 25 percent of annual rainfall may recharge the aquifer. In contrast, in desert regions recharge is very small, perhaps only 1 percent of rainfall or less. Aquifers in these areas may contain very old water, which has accumulated over centuries or under different climatic conditions.

Although commonly displayed on a two-dimensional surface, hydraulic-head distribution is generally a three-dimensional phenomenon (Figure 1.3). That is, hydraulic head varies vertically as well as laterally. The vertical distribution of hydraulic head can be determined by drilling wells near each other (nested) in the same vicinity but open to different depths. If hydraulic head increases with increasing depth, groundwater flow is upward and, in general, indicates an area of discharge. If hydraulic head decreases at a given location with increasing depth within the aquifer, then groundwater flow is downward, indicating an area of groundwater recharge.

For a local-scale flow system as defined by Toth (1998), the distance between recharge and discharge points is relatively short, and groundwater travel times through the system may be on the order of days to years. For regional-scale flow systems, the distance between recharge and discharge points is much greater, and travel

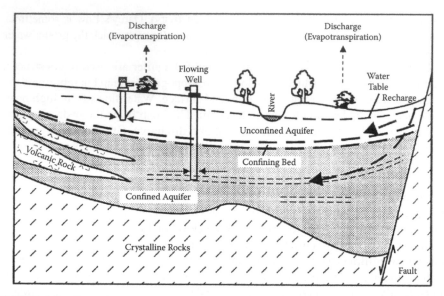

FIGURE 1.3 Groundwater movement. (From Moore, J., *Field Hydrology: A Guide for Site Investigations and Report Preparation*, CRC Press, Boca Raton, 2002. With permission from Taylor & Francis.)

times for groundwater are also much greater, on the order of decades to centuries. The determination of groundwater flow rates and directions is frequently difficult because of local variations in hydraulic conductivity, changes in stream stage, well withdrawals, and rainfall, just to name a few factors.

Many surface-water bodies (lakes, rivers, springs, and seeps) are *outcrops* of the water table, similar to the outcrops of rocks. Especially in humid regions, these surface-water bodies are useful for inferring water-table elevations where no wells exist. A stream into which groundwater naturally discharges is called a gaining stream. In arid regions, streams commonly lie above the water table, and water seeping downward toward the stream bottom provides recharge to the underlying groundwater system; these streams are called losing streams. A spring is a place of natural groundwater discharge from a natural opening at the land surface. Springs may be classified according to the geologic formation. They may also be classified according to the amount of water they discharge (large or small), their water temperature (thermal, warm, or cold), or the forces from which they are formed (gravity or artesian pressure). A spring is the result of an aquifer being filled to the point that the water overflows onto the land surface. Springs may also flow from fractures (or other openings) connected to confined aquifers.

Withdrawing groundwater from shallow aquifers that are connected to streams can have a significant effect on streams. The withdrawal can diminish the surface-water supply by capturing some of the groundwater flow that otherwise would have discharged to the stream or by inducing flow from the stream to the aquifer. This change in direction of flow from the stream to the well also can result in the movement of contaminants from the stream into the groundwater system and ultimately into the well.

1.5 SOURCE OF WATER TO A WELL

C. V. Theis (1940) described the response of an aquifer to withdrawal from wells as follows (Figure 1.4):

1. The groundwater system before development is in a state of dynamic equilibrium: recharge to the aquifer is equal to groundwater discharge.
2. When the pump is first turned on, the source of water to the well is groundwater storage.
3. At a later time, water is obtained from groundwater that would have discharged to a stream.
4. At a much later time, another source of water to the well is the result of an increase in recharge due to lowering of water level in the recharge area (increase in unsaturated zone).

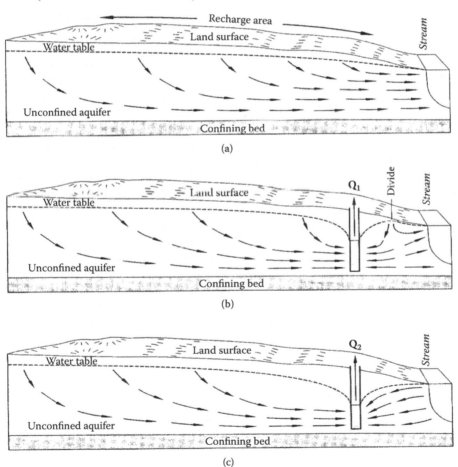

FIGURE 1.4 Source of water to a well. (From Moore, J., *Field Hydrology: A Guide for Site Investigations and Report Preparation*, CRC Press, Boca Raton, 2002. With permission from Taylor & Francis.)

1.6 LOCATING GROUNDWATER

Several techniques can be used to locate groundwater and to determine depth to water, yield, and quality. The topography may offer clues to the hydrogeologist about the occurrence of shallow groundwater.

Conditions for large quantities of shallow groundwater are more favorable under valleys than under hills. In parts of the arid southwestern United States, the presence of "water-loving plants" (phreatophytes) such as salt cedar and cottonwood trees indicates the presence of shallow groundwater. Areas with springs, wetlands, and seeps also indicate the presence of groundwater at shallow depths (Heath, 1983).

Rocks are the most valuable clues to the potential presence and availability of groundwater. The hydrogeologist should prepare maps and sections showing the distribution of the rocks both on the surface and underground.

The hydrogeologist obtains information on the wells in the area: their locations, depth to water, well yield, and geology of rocks penetrated by wells.

Wells that fail are typically too shallow or constructed in porous media with small effective porosity. These wells are generally not deep enough to provide adequate storage of water. If usable water is available at a greater depth, then the well can be deepened. Often, wells that are installed in bedrock cannot easily be deepened. Well failure may also be due to poor well design or construction or to incorrectly sized or blocked screen.

The types and orientation of joints and other fractures may be clues to obtaining useful amounts of groundwater and indicate preferential paths of movement of water and any contaminants in the water. A fracture that has a thin coating of soil can be identified as a zone of different colored vegetation, or by a slight indentation on the surface. A field technique that has gained favor by hydrogeologists is fracture-trace analysis. Fractures are found in many different rock types. Fracture traces are identified by a study of linear features on aerial photographs (Fetter 2001). Natural linear features on aerial photographs are seen as tonal variation in soils, vegetative patterns, straight stream segments or valleys, aligned surface, or other linear features. Straight stream reaches or surface segments can be visible on the surface. Rows of trees may be aligned in a floodplain. Fracture traces are the surface expression of faults, or joints. Springs may indicate location of fracture trace. Caution should be exercised, however, because in some cases the traces may be the result of deep-seated man-made structures.

1.7 GROUNDWATER FACTS

Groundwater is one of the nation's most valuable natural resources. Groundwater in the United States is being increasingly developed for irrigation and municipal supplies. The use of groundwater will likely increase in the future as surface-water supplies are diminished (in quality and quantity) and reservoir space is used up. Some key facts about groundwater follow:

1. Groundwater supplies more than half of the drinking water consumed in the United States and 96 percent of the drinking water consumed in rural areas of the nation.
2. The Ogallala aquifer (High Plains aquifer), which extends from Nebraska to Texas, supplies 30 percent of the groundwater used in the United States.
3. Groundwater withdrawal has caused the lowering (subsidence) of the ground surface by as much as 3 meters in the Houston–Galveston area of Texas, resulting in coastal and inland flooding.
4. Land in the San Joaquin Valley of California has sunk as much as 8 meters since the 1920s as a result of groundwater withdrawal.
5. Mining of groundwater (withdrawing more groundwater than is being recharged) in the northern Midwest has caused water levels in some areas to decline by as much as 300 meters.
6. In the Chicago area, the switch in water supply sources from groundwater to surface water from Lake Michigan has caused the groundwater level to rise by as much as 300 meters.
7. Groundwater is a major contributor to the flow in many streams and rivers and has a strong influence on river and wetland habitats for plants and animals.
8. As surface water becomes fully developed and appropriated, groundwater offers the only available source for new development.
9. Groundwater occurs almost everywhere beneath the land surface.
10. Movement of groundwater is very slow. A velocity of 1 meter per day is considered to be high.

ADDITIONAL RESOURCES

Carrillo-Rivera, J. J., and Cardona, A., 2008, Groundwater flow system response in thick aquifer units: theory and practice in Mexico. Selected Papers, XXXIII IAH International Congress, Zacatecas, México. International Association of Hydrogeologists, Balkema. Taylor & Francis Group, 25–46 pp.

Fetter, C. W., 2001, *Applied hydrogeology*, fourth edition. Upper Saddle River Cliffs, NJ: Prentice Hall, 598 pp.

Freeze, R. A., and Cherry, J. A., 1979, *Groundwater*. Englewood, NJ: Prentice Hall, 604 pp.

Heath, R. C., 1983. *Basic ground-water hydrology*, U.S. Geological Survey Water-Supply Paper 2220, 235 pp.

Hutson, S. et al., 2004, *Water use in the United States in 2000*, U.S. Geological Survey Circular 1268, 70 pp.

Moore, J. E. et al., 1995, *Groundwater: A primer*, Alexandria, VA: American Geological Institute, 52 pp.

Theis, C. V., 1940, The source of water derived from wells: essential factors controlling the response of an aquifer to development, *Civil Engineering*, vol. 10, no. 5, 277–280 pp.

Tóth, J., 1998, Groundwater as a geological agent, *Hydrogeology Journal*, vol. 7, 1–14 pp.

2 History of Hydrogeology in the United States

2.1 BACKGROUND

The United States has a rich history of hydrogeology. Prior to 1900, hydrogeologic studies were carried out chiefly by geologists who were self-taught in groundwater science.

From 1900 to 1930, the principal interest in hydrogeology was in areal resource investigations and the development of the underpinnings of the science. By the 1930s, agricultural irrigation was becoming more important to the economy of farming areas, and the need for more quantitative analyses and predictions drove hydrogeology in a more quantitative direction—with the focus on well hydraulics—while simultaneously maintaining the focus on hydrogeological resource appraisals for water supply. The 1960s saw a shift from well hydraulics to regional aquifer system analysis as analog model technology advanced, which then was replaced in the 1970s by digital computer model technology. In the 1970s, emphasis in hydrogeology slowly shifted from issues related to water resources to issues related to contaminant hydrogeology. In the 1980s, societal recognition of groundwater contamination as a serious environmental problem led to a major increase in the employment of hydrogeologists and advancements in the science and understanding of solute-transport phenomena and multiphase flow and transport of organic chemicals. A 1932 photograph of the U.S. Geological Survey (USGS) Office of Groundwater is shown in Figure 2.1.

2.2 MILESTONES IN THE HISTORY OF HYDROGEOLOGY IN THE UNITED STATES (1879–1988)

Major parts of the following time line were contributed by Dr. Joseph S. Rosenshein (USGS, retired).

1879—U.S. Geological Survey is established. The objective is to apply science and to publish results of studies promptly.

1885—T. C. Chamberlin's classic report on artesian wells is published.

1896—The first USGS Water-Supply Paper, *Pumping Water for Irrigation*, G. K. Gilbert's report on the Arkansas Valley of Colorado, is published.

1896—N. H. Darton reports on the Dakota Sandstone in South Dakota, a classic artesian aquifer system. He describes vertical leakage through low-permeability shales.

1899—P. B. King's classic report on groundwater principles is published.

FIGURE 2.1   USGS Division of Groundwater 1932.

1906—A. C. Veatch and C. S. Slichter complete one of the early quantitative groundwater investigations. Veatch showed that groundwater levels in an artesian aquifer respond to the influence of the elastic deformation of the aquifer by superimposed loads such as railroad trains and ocean tides.

1907—Deep well turbine pumps are developed for irrigation in California.

1909—E. E. Ellis completes a study of occurrence of water in crystalline rock in Connecticut.

1912—O. E. Meinzer becomes chief of what would eventually be designated as the Ground Water Branch of the USGS. He is considered the father of modern-day groundwater hydrology.

1912—C. H. Lee publishes the results of a study of the Owens Valley, California, which included tank experiments of rates of water use.

1916—W. C. Mendenhall publishes a report on the San Joaquin Valley, California.

1923—O. E. Meinzer reports the *Occurrence of Ground-Water in the United States* with a discussion of principles.

1925—G. S. Brown studies saltwater encroachment along the Atlantic Coast.

1928—O. E. Meinzer's paper on the compressibility of artesian aquifers is published.

1932—O. E. Meinzer prepares a paper outlining methods to estimate ground-water supplies.

1935—C. V. Theis (Figure 2.2) publishes an equation to describe nonsteady groundwater flow to a well. It is considered the greatest single contribution to the science of groundwater in this century.

1935—V. Stringfield prepares a potentiometric map of Florida, one of the first maps prepared in the United States.

1937—C. F. Tolman publishes a groundwater textbook, *Ground Water* (McGraw Hill).

1939—O. E. Meinzer publishes a report, *Ground Water in the United States.*

1940—C. V. Theis publishes a paper on the source of water to wells.

1940—K. Hubbert publishes a report on the general theory of groundwater motion.

1941—C. V. Theis presents a method to estimate the effect of groundwater withdrawal from a well on a nearby stream.

1945—C. E. Jacob develops an analytic solution for recession of the potentiometric surface when recharge ceases.

1946—C. E. Jacob develops a physical basis for the Theis equation.

1947—Carslaw and Jaeger publish a classic book on conduction of heat in a solid.

1948—J. Ferris applies the Theis equation and the method of images for locating hydrogeologic boundaries.

1950—M. D. Foster demonstrates by laboratory experiments that the occurrences of high bicarbonate groundwater in the Atlantic Coastal Plain are caused by ion exchange.

1951—H. E. Thomas publishes a book, *The Conservation of Ground Water.*

FIGURE 2.2 Photo of C. V. Theis, 1939. (From Water-Supply Paper 2415, United States Geological Survey.)

1952—R. Bennett completes a report on the Baltimore area presenting an early use of flow net analysis.

1954—H. Skibitzke develops an electrical analog model to solve a groundwater flow problem.

1955—M. Hantush and C. E. Jacob develop mathematical models to analyze leaky aquifer problems.

1959—H. Cooper reports on the dynamic balance of freshwater and saltwater in a coastal aquifer.

1959—J. Hem publishers his classic paper "Study and Interpretation of the Chemical Characteristics of Natural Waters."

1959—D. Todd publishes his book *Ground Water Hydrology.*

1960—Garrels publishes his text on mineral equilibria.

1960—M. Hantush publishes his modified artesian aquifer equation to include storage in the confining layer.

1962—J. Ferris publishes *Theory of Aquifer Tests*, a USGS Water-Supply Paper.

1963—McGuiness prepares a report on groundwater conditions in each state.

1963—R. W. Stallman points out that groundwater flow is an efficient mechanism for the transport of heat.

1963—J. Toth receives the first Meinzer Award from the Geological Society of America for his paper on regional groundwater flow.

1963—H. Cooper and Rorbaugh develop methodology for estimating changes in bank storage and groundwater contribution to streamflow.

1963—B. Walton and Prickett prepare one of the first papers demonstrating the use of analog models in hydrogeology.

1964—V. T. Chow publishes *Advances in Hydroscience.*

1965—H. LeGrand publishes a paper on groundwater contamination.

1965—P. Domenico and Mifflin publish a paper describing an analytical method to evaluate land subsidence.

1966—W. Back develops a geochemical interpretation of groundwater flow patterns from hydrochemical facies.

1967—Freeze and Witherspoon publish a numerical solution of groundwater flow in complex hydrogeologic environments.

1968—Pinder and Bredehoeft publish a paper describing the use of a digital model in aquifer evaluation.

1969—J. Poland and Davis prepare a report on subsidence due to fluid withdrawal.

1969—P. Neuman and Witherspoon publish a paper that summarizes various methods for solutions to leaky aquifer situations.

1970—J. Bredehoeft and Young prepare an article that demonstrates the interrelation of groundwater, surface water, and water use.

1973—Pinder and Bredehoeft develop a numerical solute transport model.

1974—A. Zohdy and others publish a technical manual on application of surface geophysics to groundwater.

1979—Freeze and Cherry publish a textbook, *Ground Water.* Englewood, NJ: Prentice Hall, 604p.

1980—Fetter publishes a textbook, *Applied Hydrogeology.*

1988—McDonald and Harbaugh develop the Mod Flow digital groundwater model. *HSGS TWRI,* Book 6, Chapter A-1.

ADDITIONAL RESOURCES

Back, W., 1960, Origin of hydrochemical facies of ground water in the Atlantic Coastal Plain, in W. Back and R.A. Freeze, Eds., (pp. 79–87), Chemical Hydrogeology, Benchmark Papers in Geology, 73, Stroudsburg, Pennsylvania: Hutchinson Ross Publishing Company.

Bennett, R. R., and Meyer, R. R., 1952, Geology and ground water resources of the Baltimore area, Maryland Department of Geology, *Mines and Water Resources Bulletin,* 4, 573 pp.

Bredehoeft, J. D., and Young, R. A., 1970, The temporal allocation groundwater simulation approach, *Water Resources Research,* 6, 1, 3–21 pp.

Brown, J. S., 1925, A study of coastal ground water, with special reference to Connecticut: U.S. Geological Survey Water Supply Paper 537, 101 pp.

Carslaw, H. S., and Jaeger, J. C., 1947, *Conduction of heat in solids,* Oxford Press, 510 pp.

Chow, V. T. Ed., 1964, *Advances in Hydroscience,* v. 1, New York: Academic Press, 640 pp.

Cooper, H. H., Jr., 1959, A hypothesis concerning the dynamic balance of fresh water and sea water in a coastal aquifer, *Journal of Geophysical Research,* 64, 461–467 pp.

Cooper, H. H., and Rorabaugh, M. I., 1963, Ground-water measurements and bank storage due to flood stages in surface streams, U.S. Geological Survey Water-Supply Paper 1536–J, 343–346 pp.

Domenico, P. A., and Mifflin, M. D., 1965, Water from low-permeability sediments and land subsidence, *Water Resources Research,* 1, 562–576 pp.

Ellis, E. E., 1909, A study of the occurrence of water in crystalline rocks, in Gregory, H. E., Underground water resources of Connecticut, U. S. Geological Survey Water-Supply Paper 232, 200 pp.

Ferris, J. G., 1948, Ground Water, in C. Wisler and E. Brater, (127–191 pp), *Hydrology,* New York: Wiley and Sons.

Ferris J. G., Knowles, D. B., Brown, R. H., and Stallman, R. W., 1962, Theory of aquifer tests, U.S. Geological Survey Water-Supply Paper 1536-E, 174 pp.

Foster, M. D., 1950, The origin of high sodium bicarbonate waters in the Atlantic and Gulf Coastal Plains, *Geochimica Acta.,* 1, 1, 33–48 pp.

Freeze, R. A., and Witherspoon, P. A., 1967, Theoretical analysis of regional groundwater flow, *Water Resources Research,* 3, 623–634 pp.

Garrels, H. M., 1960, *Mineral equilibria at low temperature and pressure,* New York: Harper Brothers, 254 pp.

Hantush M. S. and Jacob C. E., 1955, Nonsteady radial flow in an infinite leaky aquifer, *American Geophysical Union Transactions,* 36, 1, 95–100 pp.

Hem, J. D., 1959, Study and interpretation of the chemical characteristics of natural water, U.S. Geological Survey Water- Supply Paper, 1473, 269 pp.

Jacob, C. E., 1946, Radial flow in a leaky artesian aquifer, *American Geophysical Union Transactions,* 27, 198–208 pp.

Lee, C. H., 1912, An intensive study of the water resources of a part of Owens Valley, California, U. S. Geological Survey Water-Supply Paper 294, 135 pp.

McGuinness, C. L., 1963, The role of ground water in the national water situation, U.S. Geological Survey Water-Supply Paper 1800, 1121 pp.

Meinzer, O. E., 1928, Compression and elasticity of artesian aquifers, *Journal of Economic Geology,* 23, 3, 263–291 pp.

Meinzer, O. E., 1932, Outline of methods for estimating ground-water supplies: U.S. Geological Survey Water-Supply Paper 638-C, 99–144 pp.

Meinzer, O. E., 1939, Ground water in the United States, a summary of ground-water conditions and resources, utilization of water from wells and springs, methods of scientific investigations, and literature relating to the subject, U.S. Geological Survey Water-Supply Paper 836-D.

Mendenhall, W. C., Dole, R. B., and Stabler, H., 1916, Ground water in the San Joaquin Valley, California, U.S. Geological Survey Water-Supply Paper 398, 310 pp.

Neuman, S. P., and Witherspoon, P. A., 1969, Applicability of current theories of flow in leaky aquifers, *Water Resources Research,* 5, 817–829 pp.

Pinder, G. F., and Bredehoeft, J. D., 1968, Application of the digital computer for aquifer evaluation, *Water Resources Research,* 4, 5, 1069–1093 pp.

Poland, J. F., and Davis, G. H., 1969, Land subsidence due to withdrawal of fluids, Geological Society of America Reviews Engineering Geology, 2, 187–269 pp.

Skibitzke, H. E., 1960, Electronic computers as an aid to the analysis of hydrologic problems, Publication 52, International Association Scientific Hydrology, Commission Subterranean Waters, Gentbrugge, Belgium, 347–358 pp.

Stallman, R. W., 1963, Notes on the use of temperature data for computing ground-water velocity, in R. Bentall, Ed., Methods of collecting and interpreting ground-water data, U.S. Geological Survey Water-Supply Paper 1544–H.

Stringfield, V. T., 1936, Artesian water in the Florida Peninsula, U.S. Geological Survey Water-Supply Paper 773–C.

Theis, C. V., 1935, The relation between the lowering of the piezometric surface and the rate and duration of discharge of a well using ground-water storage, *American Geophysical Union Transactions,* 16, 519–524 pp.

Theis, C. V., 1940, The source of water derived from wells, *Civil Engineering,* 10, 5, 277–280 pp.

Thomas, H. E., 1951, *The conservation of ground water,* McGraw-Hill: New York, 327 pp.

Todd, D. K., 1959, *Ground water hydrology,* Wiley and Sons, 336 pp.

Tolman, C.F., 1937, *Groundwater,* New York: McGraw Hill.

Toth, J., 1963, A theoretical analysis of groundwater flow in small drainage basins, *Journal of Geophysical Research,* 68, 16, 4795–4812 pp.

Veatch, A. C., 1906, Fluctuations of the water level in wells, with special reference to Long Island, New York, U.S. Geological Survey Water-Supply Paper 155, 83 pp.

Walton, W. C., and Prickett, T. A., 1963, Hydrologic electric analog computer, American Society Civil Engineers Proceedings, *Hydraulics Division Journal,* 89, HY6, 67–91 pp.

Zody, H. A. R., Eaton, G. P., and Maby, D., 1974, Application of surface geophysics to ground-water investigations, *U.S. Geological Survey Techniques of Water-Resources Investigations Book 2,* Chapter D1.

3 Planning a Field Investigation

According to McGuiness (1969), the hydrogeologist's work has many facets and has changed with time. What does the hydrogeologist see as the deficiencies in knowledge that are his responsibility to remedy? At one time he had a rather simple job. He mapped the geology of his study area or refined mapping done earlier by others. He gathered as much information as he could on the depth and productivity of wells, the kinds of rocks they penetrated, and the chemical characteristics of the water they yielded. He interpreted these data in terms of the different geologic units. Ultimately, he prepared a report, maps, and charts showing where and at what depths water could be obtained, and in a general way, how much and of what quality. For rocks not penetrated by enough wells in his area to yield reliable information, he extrapolated information from other areas where similar rocks were better known, according to his experience and his familiarity with the data and published literature. This procedure was considered acceptable and sufficient in the early days of hydrogeologic studies. As groundwater development progressed and water demands became larger, the hydrogeologist found it necessary to think in quantitative terms about the permeability and storage coefficients of the aquifers and the effects on groundwater levels of pumping increasing amounts of water from more closely spaced wells.

The hydrogeologist tried to keep track of the effects of the withdrawals, not only on water levels but in diminishing natural discharge, perhaps increasing natural recharge, inducing the inflow of salt water or other water of undesirable quality, and so on. She discovered that it was not only water levels that were affected. In certain areas she found that the withdrawal of groundwater actually caused the land surface to subside.

Gradually, she began to realize that she was dealing with a system—a geologic-hydrologic-chemical-biologic-societal system in which substances and events of nature and the actions of humanity interacted so complexly that only by understanding how the system operated could she predict its response to future events of nature and actions of humanity and, thus, provide a basis for controlling the response in desired ways. This is where we are today. We are trying to define subsurface hydrologic systems in terms of their internal characteristics and their external boundaries and to devise models of them that will simulate their response to various planned or otherwise anticipated events. Repeated tests of the models will reveal the fidelity of their simulation of the prototype and will reveal also their sensitivity to data inputs of various types and thus serve as a guide to data-collection programs. In addition, the models will form the physical basis of the planner's overall model of the hydrologic socioeconomic system.

What is this groundwater system whose operation the hydrogeologist must understand? It is first of all a geologic system. It consists of consolidated rocks and/or layers of unconsolidated to semiconsolidated sediments, formed and placed where they are by geologic processes. It is a hydrologic system because the rocks contain openings that can store and transmit water. It is a chemical system because the rock skeleton, far from being inert, reacts chemically with the water and its dissolved and entrained constituents, and in this reaction changes are produced in both the rocks and the water. It is a biologic system because living organisms play an important part in the chemical reactions that take place in it, or at least in determining the chemistry of the water that enters it.

3.1 PROJECT PLANNING

An orderly plan is needed to direct the field aspects of a hydrogeologic project and the preparation of a report from conception through completion. An example of the steps in a quality assurance system to guide and ensure the success of the project is shown below. This system is used in many U.S. Geological Survey (USGS) offices, is a key to good planning, and includes clearly defined project objectives.

Project planning begins with a review of the cooperator or client's needs. The needs are described in terms of project objectives and deliverables (data and reports). The project manager develops a budget and work schedule to fit these needs. When the schedule and budget are defined, the project can plan for resources, including staff and materials. It may be necessary to negotiate some of the details with management. The project manager must be clearly granted authority that matches responsibility and should be given the resources to get the job done.

To properly plan a hydrogeologic site investigation, the purpose of the investigation, general geologic and hydrologic characteristics of the site, and the management constraints for the investigation must be defined. Subsurface investigations are a dynamic and inexact science. The success of a groundwater investigation relies not only on the technical expertise of the hydrologists involved but also on the effectiveness and efficiency of project management. Groundwater investigations are based on the creation of an accurate conceptual model.

3.1.1 FOUR STAGES OF EVERY PROJECT

1. Starting the project: generating, evaluating, and framing the need for the project.
2. Organizing and preparing: developing a plan for conducting the field investigation and preparing the report.
3. Carrying out the work: establishing the project team and assigning tasks and responsibilities.
4. Closing the project: completing the report and assessing the project results.

The key to successful project management is thorough planning (Figure 3.1 versus Figure 3.2). A thorough technical understanding of the geology and hydrology of the project (or study area) is a key factor in a successful project. If the project is planned in detail before it is undertaken and if the plans are revised as necessary during the project, the project report should be relatively easy to produce and the schedule will be

---

- The project proposal includes clear objectives, adequate planning, and a detailed work plan

- Reasonable goals

- Adequate budget

- Frequent reviews

- A technically correct and readable report

- A technically capable staff

- Timely completion of report

FIGURE 3.1 Ideal project. (From Moore, J., *Field Hydrology: A Guide for Site Investigations and Report Preparation*, CRC Press, Boca Raton, 2002. With permission from Taylor & Francis.)

met. In fact, report planning is best carried out as an integral part of the initial planning phase of the project. A senior USGS hydrologist told me many times that the report should be 90 percent completed before the hydrogeologist leaves the field.

A preliminary report outline should be developed to ensure that all technical requirements are met. To ensure that there is a common understanding of the goal, the management and the client should review the outline. Questions on the objectives and scope of the project should be resolved before the fieldwork starts. In some cases this development and review of the report outline takes place during the proposal negotiations. The report content should be described in a memorandum or other document so that the client does not forget the agreement. Such a memorandum limits the scope of the project so the client cannot add additional work at the last minute.

Contingency planning is essential for the success of a project, and in some situations, the survival of the project may require such planning. An experienced project manager can anticipate areas where problems may arise or where field or laboratory

---

- Unclear objectives

- Lack of a thorough planning

- Unreasonable goals

- Inadequate supervision

- No report outline

- No project reviews

- Over-optimistic scheduling

FIGURE 3.2 Nonideal project. (From Moore, J., *Field Hydrology: A Guide for Site Investigations and Report Preparation*, CRC Press, Boca Raton, 2002. With permission from Taylor & Francis.)

results may require changes in direction. If these potential problem areas are identified before the project starts, the budget and schedule can provide for flexibility. These problems should be discussed with the client, who will not appreciate surprises of this sort at a later date (Figure 3.1).

3.1.1.1 Project Definition

A project is a series of related activities with specific objectives, a beginning date, and end date. The major elements of project planning are the project proposal, which should include a detailed work plan and a report outline (Figure 3.2). The steps that should be followed in planning and completing a project are as follows:

1. Define the project objectives.
2. Select an approach to accomplish the objectives.
3. Decide on the major milestones for the project.
4. Select dates to begin and end the milestones.
5. Determine the budget for manpower equipment, sampling, contractors (e.g., drillers and laboratory).
6. Select and assign manpower to accomplish the work.
7. Carry out the project.
8. Write and deliver a quality technical report on time.

Sound planning. Planning provides the project chief with the tools needed to design and complete the project (and report) within the allotted time and budget. A project can be successful only when the project chief has thoroughly planned all foreseeable aspects of the project before the project begins. Project objectives must be specific, deadlines and budgets must be realistic, and difficulties must be anticipated.

Project deliverables. The most important part of project planning is a thorough understanding of the scope of work, the nature of the deliverables, and the client's expectations. The best time to clarify objectives is before the project starts. Once this understanding is reached, detailed plans can be prepared for executing the project. The list of deliverables is the natural starting point of the review, because they are what the client expects to receive.

Project budget. Once the technical requirements have been determined, the budget can then be developed. In some cases, the client specifies the total project budget; in other cases, the budget is developed to fit the technical requirements. In this case, the budget is presented to the client as part of the proposal for review and approval. The detailed project budget should include itemized costs for each activity, including project management and administration. This budget then specifies the level of staff effort and other resources that can be used to complete the required work. The budget includes costs for laboratory analyses, fieldwork, travel, editorial and technical review of the report, report printing, and equipment purchase or rental.

Project schedule. The project schedule is developed from the deadlines imposed by the project manager or supervisor in consultation with client. The project manager then examines the technical plan, available manpower, and budget, and determines the amount of time that should be devoted to each activity. This schedule should be detailed enough to show all required activities and milestones, including separate periods for report writing, review, and revision. On complex projects, several tasks may be executed simultaneously. In this case, interdependencies among the tasks must be taken into account on the schedule.

Project staffing. The schedule and budget define the level of effort that can be expended to perform the project. Before any work begins, the manager must determine the types of services needed to execute the project. Any services needed, such as drilling, should be scheduled to ensure that they will be available when required. Laboratories, for example, could be alerted to expect samples for analysis. The most important resource is the manpower that will perform the work. In most offices, several projects are being carried out at the same time. This means that the projects may be in competition for the same staff. Early planning of manpower needs allows the project manager to go to his management with a clear picture of his needs for staff resources. Management can then take the steps needed to provide the resources, or at least tell the project manager of anticipated difficulties so that contingency plans or alternative arrangements can be made.

This is the ideal situation, but it can work only if all projects are well planned, the organization is not overextended, and flexibility exists for coping with unanticipated emergencies. A single project may use personnel from more than one office, and scheduling to meet the needs of each office is essential.

Report outline. The final report is the most important deliverable for the project. The report is usually the only product of a project that is seen by the client, and in most cases it determines the success or failure of the project. The final report is often incorporated as a whole or in part into other documents, such as a license. An outline of the final report should be prepared before the project starts. The scope of the work and other client technical requirements are the basis for the final report, so preparing an outline is straightforward. The report outline should be thematic; that is, it should contain a sentence describing each report section. Prepare a topic sentence or introductory paragraph for each section of the report. The outline should be as detailed as possible. The outline is also useful in testing and in the manager's understanding of the project. The entire report can be treated this way, with the exception of details of the findings and recommendations. If several similar projects are to be performed, it may be efficient to use a standard report format that may vary only in recommendations, data, and hydrologic description. Chapter 8 goes into more detail on report.

3.1.2   TYPES OF PROJECTS

The scientist or engineer may be called on to plan many types of projects. In consulting, the most common projects are short term (less than 6 months) and have a budget of less than $50,000. Clients may be federal, state, or local government agencies; insurance companies; industrial firms; or financial or legal firms and management districts.

The scientist or engineer is frequently involved in performing research under government contracts or grants. A government employee will often work on long-term projects funded by the agency or may be called on to manage work contracted to other federal agencies, private contractors, or universities.

Short-term projects. For many consultants in the earth sciences and engineering, short-term, single-purpose projects make up the bulk of the work. Such projects are often done with a very low margin of profit and consequently must be run very efficiently. These projects may take only a day or two or may last several weeks. Usually there is a single, well-defined objective. A typical short-term project will require document review, a limited amount of field and laboratory work, data interpretation and evaluation, and a final report.

Projects of this type are very useful in developing project management skills in less experienced staff members as they require careful attention to planning and monitoring of progress and expenditures. Successful completion of short-term projects is good training for managers and builds confidence. It is still important, however, for upper management to watch over the less experienced project manager and to provide guidance or support as needed. In many firms, these projects are the bread and butter of the workload, and they help to keep staff members productive between tasks on larger projects.

Long-term projects. Projects that have a single objective and that last from several months to several years present more of a management challenge. Specifying short-term tasks is the best way to run these projects, or specifying milestones that must be met during the course of the project best ensures successful project completion. The greatest danger of a long-term, single-objective project in a busy organization is that work on it will be delayed to meet short-term organizational needs, especially crises. This commonly results in the project staff operating at panic speed as the deadline for project completion nears. Projects of this type result in major problems of quality and time or cost overruns more than any other type of project.

Open-ended projects. In many respects, the open-ended project is the most difficult to manage effectively. This type of project has no fixed completion date and commonly has one or more objectives that lack clear criteria for completion. Fiscally, such projects are often run as level-of-effort projects up to a ceiling amount. The only way to be successful is to define interim objectives or milestones with an associated schedule, budget, and report. This approach has the advantage that progress can be clearly demonstrated during the course of the project, which may make it easier to obtain additional funding as the need arises. Many research projects performed at universities and government laboratories are of this open-ended type.

3.1.3 PROJECT PROPOSAL

A project proposal is a plan to solve or address a specific problem or set of problems (Figure 3.3). The proposal should outline the technical objectives of the project, the period of time needed to achieve the project objectives, and the funding necessary to complete the work. A proposal should be clear and concise and should address the what, why, where, when, and how of the project. It should follow a standard format and contain enough information to allow the client to evaluate the proposal and report plan as follows:

Title. Create a project title that relates to the purpose, scope, and location of the proposed study. Ideally, the title should closely resemble the title of the proposed principal report that will result from the study. The title should be concise yet informative.

- Title

- Need for project

- Purpose and objectives

- Technical content and extent of study

- Approach

- Project benefits

- Planned reports

- Work plan

- Personnel

- Budget

FIGURE 3.3 Project proposal. (From Moore, J., *Field Hydrology: A Guide for Site Investigations and Report Preparation*, CRC Press, Boca Raton, 2002. With permission from Taylor & Francis.)

Statement of the problem. Explain why the project deserves the commitment of time and money. The project must produce results worthy of funding. The need for the study must be greater than just the satisfaction of intellectual curiosity.

Objectives. Relate the proposed technical results to the expressed need for those results. The objective should be specific. This is one of the most important factors in evaluating the project proposal.

Approach. Describe how the objectives will be addressed. If standard approaches and methods are to be used, a brief description will suffice. If the approach is new and untested, a more detailed description will be needed.

Benefits. Show how the results of the project will be of benefit to the client or agency and/or to the science.

Reports. Describe the planned report or reports. State probable title(s) of report(s), publication outlets, and milestone dates. Important milestones include the preparation of report outlines, report writing, colleague review, submittal of the report for

- Enthusiasm

- Disillusionment

- Panic

- Search for the guilty

- Punishment of innocent

- Praise and honors for the nonparticipants

FIGURE 3.4 Six phases. (From Moore, J., *Field Hydrology: A Guide for Site Investigations and Report Preparation*, CRC Press, Boca Raton, 2002. With permission from Taylor & Francis.)

- Define the project objectives

- Select an approach to accomplish the objective

- Select milestones for project

- Define dates to begin and end milestones

- Assign manpower to do the work

- Carry out project

- Prepare a quality report on time

FIGURE 3.5 Project planning. (From Moore, J., *Field Hydrology: A Guide for Site Investigations and Report Preparation*, CRC Press, Boca Raton, 2002. With permission from Taylor & Francis.)

approval, anticipated date of approval, and delivery date. All report activities should be planned for completion by the end of project funding.

Work plan. Schedule the starting and ending dates for each work element. Remember that some elements might be performed or executed concurrently, whereas others must be completed in sequence.

Personnel. List personnel needs by specialty and time needed. Note that all personnel must be available at the time needed in the work schedule. Indicate too the possible need for outside advisors and consultants.

Project costs. With adequate reference to plans, schedule, and personnel, itemize costs for each fiscal year. Be certain that the budget is adequate for all planned project activities for the anticipated period of study, including all costs associated with publishing the report(s).

3.1.4 SUMMARY OF PROJECT PLANNING

The inclusion of benefits, work plan, and summary in project proposal greatly enhances the client's ability to focus on key issues and to grasp quickly the importance of the work proposal (Figures 3.4 and 3.5). The work plan is an essential part of project planning. The major causes of project failure are as follows:

- Poorly prepared proposal
- Nonspecific objectives and approach
- Cost cutting
- Failure to reduce project scope when funding is reduced
- Not adhering to sound principles of cost estimation

The steps in project planning are as follows:

- Define objectives
- Prepare detailed project proposal

- Prepare detailed work plan
- Write report plan

3.2   PROJECT MANAGEMENT

Management is the organizational control used to achieve the project objectives. Management begins with a well-prepared proposal and follows through until completion of all project deliverables. Without management, the project will probably exceed budget and result in a late report.

Management by objectives. Management by objectives (MBO) is a major technique used by government agencies and the private sector to define project objectives and monitor project progress. MBO was first used by Drucker (1964). It has been used successfully at General Motors, DuPont, Ford Motor Company, and General Mills, just to name a few companies. MBO defines the following: what must be done, how it must be done, how much it will cost, what constitutes satisfactory performance, how much progress is being made, and when action should be taken to revise the project objectives and schedule (Figure 3.6). Major steps in MBO system are as follows:

1. Supervisor and project manager meet to discuss the objectives and approach for the project.
2. They agree on the objectives, approach, deliverables, work plan, and report.
3. The project manager strives to accomplish the objectives agreed upon.
4. Supervisor and project manager meet on a regular schedule to evaluate accomplishments. They review milestones, revise the schedule, correct problems, evaluate report progress and funding, and plan for the next meeting.
5. The client is kept advised on project and report progress.

Project review. The major element of project management is a periodic review of progress. Written and oral reports on work progress are needed at least quarterly. Opportunities for review are staff meetings, technical seminars, and briefings for

---

- Define the project objectives

- Select an approach to accomplish the objectives

- Decide on the major milestones for project

- Assign manpower to accomplish work

- Carry out project

- Have frequent reviews of project progress

FIGURE 3.6   Management by objectives (MBO). (From Moore, J., *Field Hydrology: A Guide for Site Investigations and Report Preparation*, CRC Press, Boca Raton, 2002. With permission from Taylor & Francis.)

cooperators and clients. An essential part of the review is to compare project progress with the work plan. Emphasis should be placed on project findings, changes in scope of work, report progress, accomplishments, needs for assistance, financial status, and future or anticipated activities. Some of the advantages of regular project review are listed below:

- Project kept on time and focused on objectives.
- Need for modifying project objectives identified.
- Personnel, technical, and financial problems identified.
- Guidance and assistance for project chief provided.
- Technical quality control provided.
- Morale improved.
- Managers, supervisors, and clients educated.
- Project and report kept on schedule.

Project management file. A project management file should be established by the project chief early in the project to maintain records and to document progress on project activity and planned reports (Figure 3.7). The file should be kept current. The following items should be included in the file:

- Project proposal
- Work plans, including milestone dates
- Budget
- Topical and annotated outlines for reports
- Lists of report illustrations and tables
- List of references for bibliographic citations
- News releases

- Project proposal

- Project description

- Work plan

- Budget

- Report outline

- List of illustrations

- List of possible report reviewers

- Report drafts

- Newspaper articles

- Summaries of project reviews

FIGURE 3.7   Project files. (From Moore, J., *Field Hydrology: A Guide for Site Investigations and Report Preparation*, CRC Press, Boca Raton, 2002. With permission from Taylor & Francis.)

- Newspaper articles on project
- Review summaries
- Report drafts and review comments
- Summary of meetings with cooperator(s) or client(s)
- All pertinent correspondence
- Purchase orders
- Cooperator or client authorizations to proceed
- Copy of contracts
- Field notes and/or project notebook
- Results of laboratory analysis
- Project data

Project controls. Project controls allow the project manager to monitor progress. The controls can provide the manager with the information needed to anticipate problems and to take preventive or remedial actions. Quality assurance is a special type of control that can help to guarantee that all project work is performed to appropriate technical standards. It may also provide the documentation needed to demonstrate in courts, in licensing hearings, or in client reviews that the work was planned to be performed in accordance with professional standards, practices, and procedures. As liability questions become increasingly important, quality assurance takes on greater importance.

Projects performed for federal agencies require frequent progress reports. Many clients are beginning to require reports on complex projects to ascertain that the project stays on track. In some companies, internal management also requires this sort of reporting.

A discussion of types of project controls and management reports that are commonly required is given below. These project control techniques are very useful to the project manager, particularly on complex or long-term projects in which parallel activities are involved. The computer graphics used to assist the control progress are described.

Work schedules. Project schedules show progress as a function of time. Every project, no matter how simple, has a schedule. This includes the beginning and completion dates. Given these two dates, the time schedule for the entire project may be developed. The project manager must calculate backward from the project end date using reasonable estimates of time for each task or activity to determine intermediate schedule points. Milestone charts are schedules commonly required in proposals and in monthly or other regularly scheduled project management reports on government projects. The milestone chart shows actual progress compared with planned progress. The schedule may also track the percentage of work completed relative to the allocated resources on a task-specific basis.

A common problem in many projects is not allowing enough time for report preparation, review, revision, and production. Failure to plan for these activities is a common reason for late completion of projects. If client review and response is required before final report delivery, it is important to show this on the project schedule. After the schedule is developed, project and support staff can be alerted to expect various activities at specific times.

Manpower projections. Manpower projections are routinely made during the planning stage of projects. These are projections of the rate at which manpower will be expended in performing the work required for the project. More sophisticated versions can show manpower broken into labor grades to make cost calculations easier. Total man-hours expended is routinely compiled at least weekly in many firms; these data are readily compared to the projection. Cumulative manpower expenditure is one important piece of management information commonly required in management reports.

Cost projections. Budget can be handled in much the same way as manpower data. Projected spending rates can be prepared from the initial project budget, together with the projection of manpower expenditures, with cumulative expenditures calculated and plotted. Comparison of actual and cumulative costs with projected costs gives a good idea of the financial progress of the project.

Technical progress. Technical progress is much more difficult to quantify than numerical measures like cost or man-hours. Many management reports ask for an estimate such as *percent complete* or the equivalent. If this estimate is to be meaningful it must be independent from accumulated costs or accumulated man-hours. This somewhat necessarily qualitative measure of the amount of "work" required for a project somehow is qualified. It is a useful management tool and if used properly will alert one of a problem or looming problem that can be addressed before disaster occurs (e.g., 40 percent of resources for project has been used and technical work has progressed 10 percent—a not-too-uncommon occurrence).

One procedure that should be formally adopted in organizations performing technical work is the review of the reports or other deliverable documents. Internal review is absolutely essential to ensure technical quality and to protect against litigation and liability, and in fact, is an integral part of sound science and engineering. Several levels of types of review exist and all have their place in project management.

The lowest level is editorial review, which is a check for spelling, typographical, and syntax errors. More important is the basic technical or peer review, wherein a knowledgeable professional involved in the work or the report reviews it for technical content. A peer reviewer has to be at arm's length from a project and therefore cannot not be a part of the project team conducting the study. All data entry and calculations should be checked (and signed off by the reviewer) independently prior to colleague review. This reviewer should also check the interpretations made by the author. The next review is the management level, where the management reviews the report to ensure that contract requirements have been satisfied, that it is consistent with other work products or other information, and that the report contents do not run counter to company or agency policies. Major technical reports, covering significant research or field or laboratory investigations, may need a peer review, especially if the findings of the project are at, near, or beyond the recognized state-of-the-art. The peer review requires that recognized independent experts in the field, often drawn from academic or governmental facilities, perform a thorough review of the data, analyses, and interpretations. This level of review is not common in the consulting world but is similar to the refereeing process used for journal publications.

Progress evaluation and troubleshooting. The ability to evaluate progress and to recognize the need for remedial actions is extremely valuable to the project manager. Because very few projects run smoothly from start to finish with no problems or surprises, this is the only way to avoid disaster. Progress reporting techniques were discussed in the previous section. The present section discusses recognizing the need for corrective action and the types of corrective actions usually possible. A supervisor often measures the success of the project by his ability to recognize problems and put corrective actions into play. A variation of corrective action that is infinitely more valuable and difficult is the preventive action.

Progress evaluation requires the comparison of project accomplishments and status with the project plans. This means that progress evaluation can only be as good as the existing plans. As was discussed in earlier chapters, the project manager must prepare plans before the start of work as carefully and in as much detail as possible. These plans must include intermediate targets, or milestones, that will be met during the course of the project. The final report then is only one of the milestones. The simplest form of progress evaluation is to compare the accomplishment at each milestone with the schedule. This, however, will tell only part of the story. Equally important is an analysis of actual expenditures against projected budget or planned manpower. Meaningful analysis usually requires that the plans, including the schedule and budget, be updated whenever conditions change.

An early slip in milestone completion means that one of two things must happen: (1) the final completion date for the project must slip, or (2) the lost time will need to be made up on later milestones in the project to allow the original completion date to be met. This may require negotiations with the client or, more commonly, finding ways to shorten future project activities. This may be done by an accelerated expenditure of manpower, a reduction in scope, or both. It can easily be seen that the earliest possible recognition of the need for action will provide the broadest choice of possibilities for remediation.

The successful project chief manager needs to be directly involved with all aspects of the project to recognize problems or to anticipate them before they occur.

Monitoring progress. Reports are used to monitor project progress. Many computer programs are available to assist the project chief in both planning and monitoring progress on the project.

Progress reports have two advantages:

1. They are a tool that allows the project manager to visualize the relation, or interdependency, of various project tasks, manpower requirements, needed services, and funding.
2. They provide a concise mechanism for reporting progress to the project staff, managers, and clients.

There are four types of progress reports: milestones, bar (Gantt), PERT, and quarterly progress review.

TABLE 3.1
Sample Milestone Chart

Milestone	Jan	Mar	May	Jul	Sep	Nov
Proposal	#	o	@			
Work plan	#	o				
Report outline		#	o			
Test drilling and sampling		#		o		
Analysis of data		#		o	@	
Preparation of report		#			o	
Review of report				#	o	
Approval				#	o	
Report delivery					o	
Postmortem					o	

Start task; o *Planned completion;* @ *Actual completion.*

A method frequently used to monitor progress is the quarterly review of project progress (every 3 months). The summary is presented orally and in written form. Items that are discussed in the review are progress on milestones, difficulties, roadblocks, and plans for the next quarter. Milestone and bar graphs can be used to describe planned and actual progress. A list of action items should be prepared after each review.

Milestone graphs are very useful for analyzing and presenting project information. Tables are prepared listing milestones, starting dates, and completion dates. The report is a list of all milestones for the project, including the final report. An example milestone chart is shown in Table 3.1.

PERT is an acronym for *program evaluation and review technique.* PERT is used extensively by government agencies and the private sector. It identifies the tasks and specifies how these relate to and depend on one another. The objective of PERT is to manage large and complex projects in the easiest and most efficient manner. It is easily adapted to designing hydrologic and geologic projects. The method that is used to apply PERT to a project is as follows:

- List project tasks.
- Determine duration for each task.
- Determine available resources (manpower, equipment, and materials).
- List costs.
- Graph activities (network).
- Compute latest allowable time.
- Add "slack time."
- Determine critical path (the longest time path through the project network). Delays in a critical task will delay the completion of the project.

Bar (Gantt) charts are used to display the project tasks in graphic form over time. These are another way of viewing the same information displayed on the PERT chart.

The Gantt chart lists tasks in rows with the durations and scheduled dates shown underneath.

Project completion. When is a project complete? Is it when the final report has been submitted to the client? When the client has accepted it, or when all payment has been received? All of these may be correct in part, and perhaps all are also wrong to some extent. The good project manager will conclude each project with an analysis of the project. This should include an examination of the technical, financial, and commercial success (or failure) of the project. Each project should be a learning experience for the manager and the organization. In this way, the process should become more efficient over time, and specific areas in need of improvement can be identified. These may include modifications in the accounting system, tracking of time and material expenditures, or internal communication procedures. If the organization is unwilling to make changes or incapable of making changes, the project manager can at least take this problem area into account when planning the next project.

A requirement for all projects is that the project be completed on time and within the allocated budget. In some cases, when the project is relatively simple and aimed at a clear and final objective, add-on work is not possible. If this is the case, success may not be apparent immediately, but only after future contracts are awarded by the client. One function of the project manager is to point out additional work that could be done by his organization for the client. This must be done carefully, so as not to alienate the client by appearing to "make work" or by interpreting the original scope of work too narrowly and requesting additional money for work the client thinks is part of the existing contract.

Completing the task on time, however, is not the main criterion that brings the client back for additional work. The project chief must deliver a quality report (product) that meets the project objectives, is well organized, is comprehensive, and is technically sound.

Completion criteria. Many projects have clearly defined endpoints. The most common endpoint is the final report. In many cases, however, the project does not really end with the delivery of final report. On level-of-effort projects, a discussion is sometimes required with the client to decide on the end of the project. For the project manager in a consulting firm, the project is not really complete until the billing has been completed and the client has paid in full. In these cases, the project manager often has to follow the project for months after all technical work has been completed. In many projects, until comments have been received and answered by submitting a revised version of the report, the project is not really complete.

The clearest criterion of completion is finishing the assignments made in the statement of work. Technical work has been completed when the client has accepted all reports and other deliverables and has certified that the contract has been fulfilled. At this point, no further costs can be incurred on that project. This limitation is precisely the reason that it is necessary to get written approval or direction from the client if the original statement of work is modified during the course of contract.

The project manager has the additional responsibility of ensuring that all costs have been accumulated for the project, all are justified expenditures on the contract

act, and the billing reflects all true costs. In some firms, this last step is left to the accountant, who prepares the actual bill and sends it to the client. The project manager's responsibility may extend to collecting payment from the client. If the client does not remit in a timely fashion, the manager may be the one to remind the client of his or her responsibilities. Requiring payment on delivery of the report avoids this embarrassment.

The project manager may need to decide which, if not all, costs are to be transmitted to the client. Any costs not passed on to the client are charged against profit, so any such decision must be made very carefully and often requires approval by higher management.

Analysis after project completion. The diligent project manager will evaluate the project after its conclusion to improve skills, to better analyze project results, and to evaluate the management system. The only way it is a useful expenditure of staff resources is if the project is analyzed for management weaknesses and strengths, with the resulting understanding applied to future project management. Many times, weaknesses in the organization, such as priority setting, staff allocation, management support, inexperienced preliminary budgeting, and scheduling, are recognized as systemic problems in the analysis. A truly useful analysis, or *postmortem*, must go well beyond the stage of placing blame, as often happens.

Problems that surface during the analysis generally fall into two categories: weaknesses in planning and weaknesses in execution. Weaknesses in planning generally result from lack of sufficient planning or from the planner's inexperience. Weaknesses in execution are more often communication problems, either with the client or with the subcontractors, or results of overcommitment of staff resources or lack of skills among the technical staff.

Roles and responsibilities. Client service can be improved through better review and communication (both internal and external). Steps that can be taken to improve the communication with the client are noted below:

- The division manager or equivalent is responsible for discussing new work with clients.
- The project manager is responsible for filling out the project assignment forms and project spreadsheets.
- The designated client contact is responsible for communicating with the client regarding all aspects (technical, schedule, budget) of the project.
- It is imperative that if there is an increase in work effort, the client is informed as soon as possible.
- The project chief must not undertake tasks beyond the project scope without management and client authorization. The client contact is responsible for daily client communication.
- The client contact should notify the client regarding the scheduled arrival date of deliverables. Courtesy calls to monitor client satisfaction should be conducted.
- The project manager is responsible for completing assignments consistent with client goals. This includes being cognizant about the technical, scheduling, and budget aspects of the project.

- The project chief should not exceed budgets and schedules. The project staff should be aware of the overall schedule and project budget including allotted hours.
- If the project chief cannot stay within the budget, the chief's supervisor must be informed immediately so that an adjustment can be made in the budget or the client can be alerted.
- Contingencies should be anticipated and planned for at the inception of the project. The client does not like unpleasant surprises.

The following is a checklist that could be used for client reports and correspondence before they are transmitted:

- Should the document be labeled "Privileged and Confidential—Prepared at the Request of Counsel"?
- Does the letter require courtesy copies to anyone?
- Should the courtesy copies be blind or noted?
- Are the copies going to the correct people?
- Should the people receiving courtesy copies also receive enclosures?
- Is the letter format correct?
- Are the addressee's name and title correct?
- Is the addressee's company spelled correctly?
- Does the name of the addressee match the salutation?
- Is the date correct at the top of the letter?
- Should any of the enclosures to a letter also be on letterhead to indicate who authored the enclosure?
- If the document is a letter contract or a professional services agreement, are two originals being transmitted to the addressee? (The addressee retains one copy and the other is signed and returned.)
- If a signature stamp is used for a letter, does the imprint look authentic, not crooked or off-center?

Total quality management (TQM). TQM is an integrated management system for achieving client satisfaction. It involves all managers and employees and uses quantitative methods to improve an organization's progress. The TQM problem-solving process consists of eight steps: identify problems, select problem, analyze root cause, identify possible solutions, select solutions, test solution, implement solution, and trace effectiveness.

3.2.1 SUMMARY OF PROJECT MANAGEMENT

Groundwater investigations are designed to study the groundwater resources of a specified area and to determine cause-and-effect relations for such phenomena as reduction in aquifer storage, reduced well yield, contamination, or deterioration in quality. The boundaries of the areas may be determined arbitrarily to coincide with county or state boundaries, or they may be based on hydrologic criteria and enclose entire drainage basins or hydrologic units.

A groundwater investigation may be separated into three steps: planning, execution, and reporting.

1. Project management begins with proper project planning and succeeds with implementation of the work plan. Well-managed projects achieve objectives and produce quality reports on time.
2. Poorly managed projects exceed the project budget and result in late or weak technical reports.
3. The project manager is responsible for

   • identifying milestones and completion dates
   • defining equipment and manpower needs
   • providing detailed progress reports

4. Management by objectives keeps the project on schedule and focused on objectives.
5. A project file should be started at the beginning of the project and maintained throughout the life of the project. The file should contain the up-to-date proposal, quarterly reviews, work plan and milestones, budget, expenditures, report topic and annotated outline, and press statements.
6. Three useful management charts are milestone, PERT, and bar (Gantt).
7. Project review is important to project success and the review should place an emphasis on findings, progress of field studies, reports, needs for assistance, plans for next quarter, and report review.
8. Each project should conclude with an evaluation of the success or failure of the project.
9. The main criterion that brings repeat business is good customer relations including a high-quality report on time that meets objectives, is well organized, and is readable by the intended audience.

Every investigation should have a plan to guide the project from conception, to development, to the conclusion. The plan can be simple or complex, depending on the objective, the scope of the investigation, the funds and time available, and the people involved. Any plan should be kept as simple as possible to guide the investigation

The project plan should also be flexible to allow for unforeseen events, for windfalls of data, and for other information. In some instances, the objectives may need to be revised during the investigation. The whole project should be examined periodically to avoid errors in planning and execution of the investigation.

3.3   TYPES OF INVESTIGATIONS

The scope of an investigation may be classified, on the basis of intensity of effort, as *reconnaissance* or *comprehensive*; these terms are used merely to indicate, in a general way, the variation in the scope of investigations. A *contamination investigation* is a special type of comprehensive investigation.

A *reconnaissance investigation* is made of an area where little or nothing is known of the groundwater resources. Commonly, the objectives are to determine the general occurrence and quality of water, the importance and types of use of groundwater in the area, and the kinds and locations of existing and potential problems, including contamination. A rapid reconnaissance may be necessary to determine the need for a more thorough or particular type of investigation. The site covered by a reconnaissance investigation may be only a few square kilometers but may include several hundreds of square kilometers.

A *comprehensive investigation* commonly covers an area not larger than a few hundred square kilometers and as small as a contamination site of a few hectares or less. The objective of a comprehensive investigation is to describe the groundwater resources of the area both quantitatively and qualitatively. Such an investigation includes sufficient analysis of recognizable problems so that the necessary data can be collected and presented in usable form to those who are responsible for the actual solution of the problems. Ideally, the relation of groundwater to surface water and the quality of groundwater, including changes owing to development, are determined and described so that the entire water resources of the area may be developed fully and efficiently. Rarely is it practical to meet all the above objectives. The investigation, therefore, generally is tailored to meet the objectives within the limits prescribed by time and funds. If the project involves tracer isotope analysis and hydrochemical modeling, it will take 1 to 2 years to evaluate the hydrogeology.

The length of time and amount of effort required for an investigation depend upon the size of the area, the complexity of the geologic and hydrologic environment, the amount of data readily available, and the scope of the investigation. Typical studies may last only a few weeks, months, or as long as a year, and the effort required may range from a few man-months to tens of man-years.

Groundwater problems may be classified as problems of quantity, quality, distribution, development, and conflict of interest; contamination may be the principal or a contributing factor in either classification. Rarely can the problems in an area be relegated to a single classification, but the terms are useful, nonetheless, for indicating the type of information and investigation that will be needed for full utilization of the resource.

A *contamination investigation* provides a comprehensive overview of the hydrogeology and contamination at a site and provides guidance for future investigations. Such investigations are commonly designated as Phase 1 and 2.

A *Phase 1 investigation* might consist of the following elements or activities:

1. Summarize the available literature
2. Provide as complete a picture as possible of the basic geologic and hydrogeologic environment on the basis of

    • literature review
    • well inventory
    • site reconnaissance visits and interviews
    • conceptual hydrogeologic model

3. Preparation of Phase 1 report

The literature review may identify concerns such as presence of faults, preexisting land uses, potential to reverse groundwater gradient, and the variability of background water quality. It would be helpful to take pictures at the site with emphasis on plant operations and the storage of waste. The Phase 1 report should give recommendations for the placement of Phase 2 monitor wells and information about the sampling plan.

A *Phase 2 investigation* would provide a detailed description of the site to test the conceptual hydrologic model. Wells are installed and water levels measured. Water-table contour maps and cross-sections are prepared. Monitor well screen intervals are selected (establish well depths and locations). Aquifer tests are run to determine hydraulic conductivity and specific yield.

A list of actions typically taken at hazardous waste sites in the early 1980s versus the actions or activities that could be recommended as of 2010 is shown below. Two data gaps are the vertical distribution of hydraulic head and hydraulic conductivity values.

Actions taken in the 1980s:

- Install a few dozen shallow monitor wells. Sample groundwater numerous times for pollutants.
- Define geology primarily by driller's logs and cuttings.
- Possibly obtain soil and core samples for chemical analysis.

Recommended actions in 2010:

- Conduct surface and borehole geophysical surveys (electromagnetic and ground-penetrating radar) as needed.
- Install depth-specific clusters of monitor wells.
- Initially sample for suite of suspended contaminants.
- Sample for aquifer hydrogeochemistry (redox conditions) to evaluate chlorinated solvent contaminants or some metals.
- Define geology by extensive coring and sediment sampling.
- Evaluate local hydrology with well clusters. Perform limited tests on sediment samples.

The results of laboratory analyses are only as reliable as the samples, field standards, and blanks received. Maintaining representative samples requires consideration of well purging, sample collection, and sample preservation. For detailed information on groundwater data collection see Koterba and others (1995) and Lapham and others (1997).

3.4   OBJECTIVES OF INVESTIGATIONS

The objectives of a comprehensive investigation include a quantitative description of the groundwater resources, their quality, and their relationship to surface water. The objectives may be stated more specifically:

1. To determine the hydraulic properties and the dimensions of each unit in the geologic section, at least down to the deepest source of water usable for any practical purpose

2. To determine the source and amount of inflow (lateral as well as vertical from above and beneath) recharge to each aquifer
3. To determine the amount and location of discharge from each aquifer
4. To determine the quality of the water from each aquifer
5. To determine the effects of withdrawal of water from each aquifer
6. To determine the effects on surface water of changes in recharge and discharge of groundwater
7. To determine the direction of water movement
8. To determine the effects on groundwater of any changes in flow and levels of surface water.
9. To determine the sources, rate and direction of movement, and fate of contaminants

3.5 SOURCES OF HYDROLOGIC DATA

The first step in a project is the collection of data and a thorough search of the literature and other sources of information. For example, much information may be stored in the files of public agencies. A search of the literature will commonly yield considerable geologic information and may give information on well drilling and water development in the area. In addition to scientific reports, newspaper files and historical articles or books may yield considerable and valuable information. An examination of surface-water records may reveal areas and magnitudes of potential recharge or discharge of groundwater.

A primary consideration in collecting data for a hydrologic investigation is to obtain all easily obtainable published materials. The first step is to obtain information on the project locally from federal, state, and university sources. The search could start with the USGS and state literature. Useful publications are hydrologic atlases, geologic maps, water supply papers, basic data reports, state (USGS) circulars, Open File Reports, and Water Resources Investigation Reports. Contact the USGS district office to see if current investigations are being or have been carried out in the locality. Literature searches by subject and area can now be made easily using computer retrievals. The library staffs of the USGS, Environmental Protection Agency (EPA), or Bureau of Land Management (BLM) can assist you in making such searches. The following are potential data sources for hydrogeologic studies:

Federal agencies: USGS, EPA, Bureau of Reclamation, National Park Service, BLM, Corps of Engineers, Soil Conservation Service (now the National Resources Conservation Service), Fish and Wildlife Service, and the Forest Service.

State and local agencies: Local city and county planning boards, state EPAs, state geological surveys, state engineer.

Knowledgeable individuals: Past well owners, federal agency personnel, local well drillers, consulting engineers, plant managers, neighbors, owners, well owners.

State and federal projects: Site-specific assessment data for dams, harbors, river basin impoundments, and highways.

American Geological Institute: Directory of Geosciences Departments, GEOREF (bibliographic database).

Government and university libraries: USGS libraries (Reston, Virginia; Menlo Park, California; and Denver, Colorado); USGS Home Page (Usgs. gov); EPA libraries in Washington, D.C., 10 regional offices, and the research offices (Cincinnati, Ohio; Athens, Georgia; Las Vegas, Nevada; and Adam, Oklahoma).

Computerized online databases: DIALOG (800-dialog) accesses over 420 databases from many disciplines including GEOREF, provides book reviews and biographies and access to newspapers, journals, and other sources; Earth Science Data Directory (ESDD) at USGS (703-648-7112) is a database that contains geologic, hydrologic, cartographic, and biological information; real-time hydrologic data are available for each state and website for each USGS district office.

Topographic data sources: USGS Branch of Distribution and Earth Science Distribution Center (800-USA-MAPS) have map data in both graphic and digital form. Many topographic maps are available online.

Aerial photographs: Earth Resources Observation System (EROS) data center USGS Sioux Falls, South Dakota (605-594-6151), has aerial photography obtained by the USGS and other federal agencies and LANDSTAT satellite imagery. Mosaic and aerial photographs can also be obtained from the USGS map sales office in Denver.

Soils: USDA Soil Conservation Service has county-level soil surveys available for about 75 percent of the country.

Geophysical maps: USGS in Denver has aeromagnetic maps, maps of magnetic delineation, landslide information, and earthquake data.

Hydrologic Information Reports: Water Data Information Coordination Program (703-648-6810) of the USGS in Reston has publications on recommended methods for water data acquisition and guidelines for determining flood frequency.

National Water Data Storage and Retrieval System (WATSTORE): USGS maintains WATSTORE for surface-water and water-quality files, and it also has the Office of Water Resources, which publishes water-resource abstracts from throughout the world.

Water Resource Discipline Offices: All 50 states have state-level water resources investigation reports and data on groundwater, surface water, and water quality; USGS publications include Water Supply Papers, Hydrologic Atlas Reports, Groundwater Atlases, Professional Papers, and fact sheets; State Geological Surveys; State Water Resource agencies have well logs, well permits, stream flow records, drilling permits; National Groundwater Information of the National Water Resources Association (Dublin, Ohio) has computerized online bibliographic database.

Climatic data: National Climatic Data Center (Asheville, North Carolina) collects and catalogs nearly all U.S. weather records.

3.6 HYDROLOGIC WEBSITES

The following is a partial listing of key hydrogeologic information websites:

Aerial photographs and maps: www.aerotoplia.com
American Geological Institute: www.agiweb.org
American Geophysical Union: www.agu.org
American Institute of Hydrology: www.aihydrology.org
American Society for Testing and Materials: www.astm.org
American Water Resources Association: www.awra.org
Environmental Protection Agency: www.epa.gov
Geoscience Information Society: www.geoinfo.org
Government Printing Office: www.gpo.gov
Groundwater information: www.groundwater.com
International Association of Hydrogeologists: www.iah.org
Merriam-Webster's Dictionary: www.m-w.com
National Climatic Data Center: www.ncdc.noaa.gov
U.S. Army Corps of Engineers: www.usace.army.mil
USGS: www.usgs.gov

3.7 GEOGRAPHIC INFORMATION SYSTEMS

A computer program called the Geographic Information System (GIS) can be used to store large amounts of data from a project. GIS can be used to store, analyze, and display geographic information in multiple layers. The program is based on maps of different features: topography, water table, well locations, location of buildings, and hydrologic features.

3.8 RESPONSIBILITY OF HYDROGEOLOGISTS

3.8.1 RULES FOR PROFESSIONAL CONDUCT

According to the American Institute of Hydrology (Registry, 2001), hydrogeologists should

1. Seek and engage in such professional work or assignments for which they are qualified by education, training, or experience.
2. Explain their work and merit modestly, and avoid any act tending to promote their own interests at the expense of the profession.
3. Avoid any act that may diminish public confidence in their profession and conduct themselves to maintain their reputation for professional integrity.
4. Be objective in professional reports and testimonies and not disseminate untrue, sensational, or exaggerated statements regarding their hydrogeologic work.

5. Avoid accepting a commission whereby duty to their client or to the public would conflict with their personal interest or the interest of another client.
6. Not accept compensation for services on the same project from more than one party unless the circumstances are fully disclosed and agreed to.
7. Give proper credit for work done by others, not accept credit due to others, and not accept employment that would replace another professional except with that person's knowledge and concurrence.
8. Associate only with reputable persons and organizations.
9. Conduct field investigations in accordance with the scientific method, collect and interpret hydrogeologic data using approved or reproducible methods, keep records of the investigations, and document the results of the investigation.

3.9   FIELD NOTEBOOK

A hydrogeologic investigation can be no better than the quality of the field data and the hydrogeologist's application of the scientific method (Fetter, 2001). It is essential that the field hydrogeologist keep a neat and complete field notebook. The field notes should be entered in the field at the time they are collected. Entries in the notebook should be written in waterproof ink in a neat, legible hand so that someone else can read it. It should contain carefully drawn sketches and photographs. You should write with a waterproof pen. The notes should never be erased. The notebook should contain the hydrogeologist's name, company affiliation, and contact information (address, email address, and phone number). The basic information that should be in notebook is the location, purpose of the investigation, and objective of each day's work. Other important information to include or describe in the field notebook is the methods used to collect the data, the serial numbers of any equipment used, calibration information for the equipment, and comments concerning field conditions or contacts required for access to property or wells.

3.10   FIELD SAFETY

Planning for Field Safety by the American Geological Institute (1992) describes the most common hazards and pitfalls hydrogeologists might encounter and suggests ways to avoid them. The book contains chapters on pretrip planning, equipment precautions, safety in the field, transportation, weather cautions, animals and plants, hazards of specific regions, and what to do in case of an emergency. Safety is a major concern when working with ground-water well drilling in remote areas.

Some of the major safety concerns for hydrogeologists are as follows:

1. A site safety plan for each project or field operation must be prepared by a qualified safety specialist for each project. The plan must be periodically reviewed to keep it current.
2. The potential weather conditions at the field site should be thoroughly investigated as part of pretrip planning.

3. Choose clothing you will need for these conditions.
4. Leave a copy of daily travel plans and campsites with your supervisor, or with a family member if self-employed.
5. Always wear eye protection when using rock hammers.
6. Drilling machines are inherently dangerous because the rotating drill stem can break bones and cause other serious physical damage. Hardhats and steel-toed boots should be worn at all times.
7. The drilling site should be evaluated for safety. Always check for overhead power lines and unstable or steep slopes.
8. Always wear a personal floatation device when in a boat or working on or near water.
9. Make sure that your vehicle is safe and in good working order before beginning each field trip, and keep a first-aid kit, emergency supplies (water, food, blankets), and tools in the vehicle.
10. Use caution when working at wells with electric pumps or other electrical equipment.

ADDITIONAL RESOURCES

American Geological Institute, 1992, *Planning for field safety,* 197 pp.

American Geological Institute, 1992, *Planning for field safety*, American Geological Institute, 197 pp.

Assaad, F., LaMoreaux, P., and Hughes, T., 2004, *Field methods for geologists and hydrogeologists,* Heidelberg, Germany: Springer, 377 pp.

ASTM, 1995, D 5753 *Guide for planning and conducting borehole geophysical logging.*

ASTM, 1996b, D 5979 *Guide for conceptualization and characterization of groundwater flow systems.*

ASTM, 1999, D 6063 *Guide to selection and methods for assessing groundwater or aquifer sensitivity and vulnerability.*

ASTM, 1996c, D 5980 *Guide for selection and documentation of existing wells for use in environmental site characterization and monitoring.*

ASTM, 1996d, D 6000 *Guide for the presentation of water-level information from groundwater sites.*

ASTM, 1994, D 5612 *Guide for quality planning and field implementation of a water quality measurement program.*

Drucker, P., 1964, *Managing for results.* New York: Harper and Row, 240 pp.

Fetter, C. W., 2001, *Applied hydrogeology: Fourth edition,* Englewood, NJ: Prentice Hall, 598 pp.

McGuinness, E. L., 1969, *Scientific or rule-of-thumb techniques of ground-water management—which will prevail?* U.S. Geological Survey Circular 608, 8 pp.

Moore, J. E., 1991, *A guide for preparing hydrologic and geologic projects and reports,* Kendall/Hunt Publishing Co., 96 pp.

Portny, S. E., 2010, *Project management for dummies,* third edition, Wiley Publishing, Inc., 264 pp.

4 Surface Investigations

4.1 CONCEPTUAL MODEL

A conceptual model of a hydrologic system is a clear, qualitative, physical description of the operation of the system. A hydrologist's conceptual model of the system determines the direction, focus, and specific content of the investigation. The model consists of maps and cross-sections showing the subsurface geology, location of potential sources of recharge, flow paths, and discharge. If the conceptual model does not accurately represent the operation of the real hydrologic system, then the results of the investigation will be at best misleading and at worst grossly in error. The conceptual model should be considered as subject to change or modification as new hydrologic information becomes available during the course of the project. The steps for developing a conceptual model are as follows:

1. Determine the type and distribution of groundwater recharge areas: infiltration from precipitation and surface-water bodies, from irrigation, as well as leakage or inflow from other aquifers or adjacent areas.
2. Determine the type and distribution of discharge of groundwater to springs and seeps, to streams and lakes, by evapotranspiration, and by leakage to other aquifers, aquitards, and wells.
3. Describe the head distribution (both vertically and aerially) and flow in three directions.
4. Describe the distribution and extent of the principal aquifers and confining beds with maps and sections.
5. Estimate hydraulic properties of the aquifers (specific yield and hydraulic conductivity).
6. Evaluate the response of the system to water withdrawals pumping from wells or diversions from surface-water bodies or other stresses.

Examples of conceptual models in different hydrogeologic settings are shown in Figures 4.1, 4.2, and 4.3.

4.2 PRELIMINARY SITE RECONNAISSANCE

A field site reconnaissance is the first step in a hydrogeologic field investigation. It provides a preliminary evaluation of the hydrogeologic conditions at the site. These site visits are interesting, exciting, and challenging because they are like a detective's investigation. At the beginning of a project, many unknowns have to be determined by scientific investigation. This is especially true for evaluations at contaminated sites.

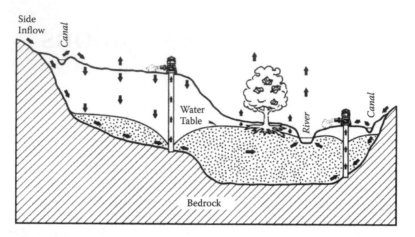

FIGURE 4.1   Unconfined aquifer section. (From Moore, J., *Field Hydrology: A Guide for Site Investigations and Report Preparation*, CRC Press, Boca Raton, 2002. With permission from Taylor & Francis.)

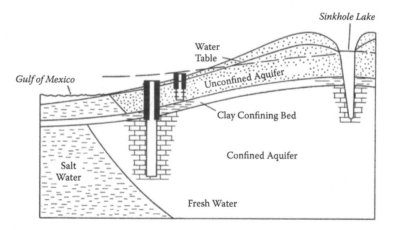

FIGURE 4.2   Confined aquifer section. (From Moore, J., *Field Hydrology: A Guide for Site Investigations and Report Preparation*, CRC Press, Boca Raton, 2002. With permission from Taylor & Francis.)

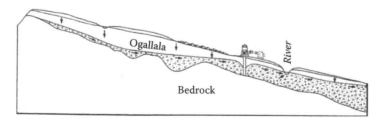

FIGURE 4.3   Unconfined aquifer section (mining). (From Moore, J., *Field Hydrology: A Guide for Site Investigations and Report Preparation*, CRC Press, Boca Raton, 2002. With permission from Taylor & Francis.)

Site investigations are carried out for almost all hydrogeologic projects. The hydrogeologic project could be an investigation for water supply, geological engineering construction, real-estate transfer, groundwater contamination, or remediation of contaminated groundwater. The planning of the field investigation will be determined by the project objectives, but many elements are common to all studies. It should be pointed out that the cost of field data collection is the most expensive part of the project. Geophysical surveys, well inventories, aquifer tests, laboratory tests, monitor wells, borehole logging, and test drilling are expensive and time-consuming. It is the responsibility of the project chief to keep these costs to the minimum required to accomplish the objectives of the project.

ASTM standards are available for investigation of soil, rock, and groundwater projects. They are intended to improve consistency of hydrologic data and to encourage organized (rational) planning of site characterization. An adequate and organized exploration program is needed because the hydrogeologic conditions at a specific site are the result of a combination of natural, geologic, topographic, and climatic factors, complicated by manmade modification.

4.2.1 Suggestions for Conducting a Preliminary Site Field Investigation

Plan and conduct a preliminary site investigation that includes geophysical surveys and simple field sampling and collection of hydrologic and other information to define and refine the conceptual model of the site. After the preliminary site investigation is completed, a detailed site investigation can be planned, if necessary.

1. Include in the site history information and maps showing where commercial and industrial activity has taken place. The site history should include location of buildings, storage tanks, transportation facilities (pipes, trucks, trains, etc.), and chemicals manufactured or used in manufacturing.
2. Study the soil and rock outcrops in the project area where this type of data is insufficient. Take pictures, make sketches, and take notes about relevant features and conditions (rocks, vegetation, drainage, man-made structures, soil, groundwater, surface water, facilities, etc.).
3. Compile a preliminary map of the project area using available topographic maps and aerial photography. Plot the location of wells, springs, streams, vegetation, and wetlands. The field site reconnaissance gives an opportunity to check the accuracy of the map that was compiled in the planning phase. It is also an opportunity to refine the conceptual model of the area. Site features that require further investigation should be identified (ASTM, 1996).
4. Select methods, dates, and frequency of field data collection periods. Accurate elevation data and x–y coordinates are critical in the data compilations and accuracy of interpretations. Develop a detailed site-sampling plan that includes description of the methods to be used to collect and analyze field data; establish protocols for sampling and field measurements and chain of custody procedures to ensure quality assurance and quality control.

5. Make preliminary geophysical surveys to select sites for additional test drilling, if necessary.
6. Identify type and extent of contaminants present, if applicable.
7. Visit municipal or county government offices to obtain valuable information on the area. Visit the local university or public library to find publications on previous studies. A trip to the local U.S. Geological Survey (USGS) district or subdistrict office can have a large technical payoff. Inquiries should be made to find out about current projects in the area and to meet with the project chief. This information may also be obtained on the USGS website for that state. Also visit the nearest U.S. Department of Agriculture (USDA) office to obtain soil maps of the study area.
8. Take preliminary water-level measurements and collect water samples, if water quality or contamination is an aspect of the project.
9. Prepare preliminary water-table contour maps and cross-section and determine groundwater flow directions.

4.2.1.1   Selection of Well-Drilling Sites

Drilling sites are selected mainly based on geologic considerations. Geologic criteria are the most important. The hydrogeologist should make the site location decision. The following information should be specified at the proposed drilling sites (Mandel and Shiftan, 1981).

1. Depth to top of aquifer—This information can be obtained or sometimes inferred from geologic maps, cross-sections, and geophysical surveys. It may be advisable to drill a small-diameter pilot well to locate the top of the aquifer and to measure the depth to the water level (see next item).
2. Water-level depth—The approximate depth to the table can be estimated from regional drainage of lakes and streams. Obtain water-level data about nearby production wells, if available.
3. Drilling depth—Estimate drilling depth from well logs and nearby well logs.
4. Stratigraphy—Identify geologic materials and changes with depth in drilling samples.
5. Potential contamination—Prepare to cement off parts of the well.
6. Nearby well interference—Estimate interference based on location, direction to and pumping from such wells, and transmissibility of the aquifer.

4.3   SITE VISIT

After the office review of historical records and documents and the title search to identify previous owners of the property are completed, the hydrogeologist returns to the site. The hydrogeologist interviews employees and neighbors who live adjacent to the property to learn about the present conditions and past history. A

FIGURE 4.4   Horizontal well.

walk-through is made to confirm the document records. The hydrologist walks the entire site, noting natural features as well as constructed ones, including depressions, canals, mounds, waste pits, and lagoons. Close attention should be given to discolored soil and any oil stains.

4.4   SPRING INVESTIGATION

Springs are points or areas of natural groundwater discharge to the land surface (Figure 4.4) and thus a source of hydrogeologic information. They occur where the water table intersects the land surface or as a surface expression of a fracture opening. Springs may be classified according to their geologic formation, the quantity of discharge, water temperature, gravity, and whether they are gravity fed or artesian.

The following information should be collected or noted when field observations are made of a spring:

- Evaluate the appearance—topographic setting, type of collection system, and extent of wetted area.
- Evaluate hydrogeologic setting, rock type, and fracturing.
- Determine the type of discharge—diffuse or single.
- Observe the types of vegetation at and surrounding the spring.
- Determine the type of soil.
- Photograph the spring area.
- Collect a water sample.
- Make field measurements of conductance, temperature, coliform bacteria.
- Measure discharge.
- Determine if the spring is protected from obvious sources of contamination.

- Literature review

- Well inventory

- Site visit

- Flow analysis

- Conceptual model

- Phase I report

FIGURE 4.5   Phase I investigation. (From Moore, J., *Field Hydrology: A Guide for Site Investigations and Report Preparation*, CRC Press, Boca Raton, 2002. With permission from Taylor & Francis.)

- Determine if the collection chamber is properly constructed.
- Determine if the spring is protected from animals.
- Determine if the site is protected from flooding.
- Determine if the located is in the recharge or discharge area.

4.4.1   HAZARDOUS WASTE SITE INVESTIGATION

Many factors influence the selection of a site for a landfill: public concerns, rainfall, runoff, and air quality. However, two of the most critical concerns are geology and hydrology. The most critical element in this type of site investigation is the development of a conceptual hydrogeologic model. The investigation can be divided into three phases: Phase I, desk studies; Phase II, field investigations, data analysis, and monitoring design and installation; and Phase III, remediation.

The Phase I investigations (Figure 4.5) are designed to provide a comprehensive overview of available information on the site. It can be conducted as the first step in a Superfund remedial investigation/feasibility study (RI/FS). Phase I provides a summary of known site features for planning. The elements of a Phase I investigation are:

- Literature review and conceptual hydrogeologic model
- Well inventory
- Site visit
- Phase I report

Information from the Phase I investigation is used to develop the conceptual model of the hydrogeologic conditions at the site. The model is the essential element for planning the field investigations in Phase II. Once the conceptual model is formulated, the need for additional data is apparent. In less complicated hydrogeologic situations, a simple installation of a few upgradient wells and several downgradient wells is sufficient. The final conceptual model points out needs for additional verification data and the desired locations of Phase II monitoring wells.

- Field investigations
  - Install piezometers
  - Measure water levels
- Data analysis
  - Prepare water table contour map
  - Define flow system
- Monitoring well design
  - Select target zones
  - Select monitor-well depth
  - Establish well locations
  - Install monitor wells
  - Test and operate system
- Phase II report

FIGURE 4.6 Phase II investigation. (From Moore, J., *Field Hydrology: A Guide for Site Investigations and Report Preparation*, CRC Press, Boca Raton, 2002. With permission from Taylor & Francis.)

Phase II field investigation (Figure 4.6) consists of the following components: detailed field data collection and measurements, data analysis, and monitoring well installation. The field investigations have the following elements:

- Map and describe site lithology and geologic structure.
- Install piezometers.
- Measure water levels.

The data analysis aspect has the following elements:

- Refine conceptual model.
- Prepare water-table contour map.
- Analyze and refine description of flow system.

The monitoring system design phase consists of the following:

- Select target monitoring zones.
- Select monitor-well depth.
- Establish well locations.
- Install monitor wells.
- Test and operate system.

Phase III consists of developing and carrying out a remediation plan that is appropriate for the problem, effective in removal or containment of contaminants, and cost-effective. Monitoring of the site must continue to evaluate the effectiveness of the remediation scheme.

4.5 GEOPHYSICAL SURVEYS

Geophysical surveys are useful to assist in the cost-effective placement of test holes, pumping wells, and monitor wells. Geophysical methods are essential at most contaminated sites to determine the extent of contamination and to avoid hazardous drilling locations. Geophysical methods together with direct-push drilling technologies are especially valuable for ensuring that monitor wells intercept contaminant plumes and to document background uncontaminated conditions. For more detailed information, refer to Zody and others, 1974; U.S. EPA, 1978; Benson, 1983; National Water Well Association, 1985; USGS Branch of Geophysics (http://water.usgs.gov/ogw/bgas/). The various types of geophysical surveys and their applications to hydrogeologic investigations include surface geophysics and geophysical well logging.

4.5.1 SURFACE GEOPHYSICS

Electrical Resistivity: Electrical resistivity measurements can be useful for determining subsurface characteristics for highway construction and dam site location. Earth resistivity methods provide assistance in distinguishing subsurface materials without physical excavation. The resistivity might help characterize the material as clay, silt, sand, gravel, or consolidated rock. The moisture content of the rock and the specific conductance of the groundwater will cause variations in the electrical resistivity of the rock. This technique is useful for

*   Subsurface stratigraphy
*   Mapping contaminant plumes
*   Locating abandoned wells
*   Mapping fresh water/ salt water
*   Indicating fracture orientation

Electromagnetic Induction

*   Mapping conductive organic contaminant
*   Locating buried utilities, tanks, and drums
*   Locating abandoned wells
*   Subsurface stratigraphic profiling
*   Locating freshwater–saltwater interface

Seismic Refraction and Reflection (Haeni, 1988)

*   Mapping top of bedrock and buried channels
*   Measuring depth to groundwater
*   Indicating fracture orientation
*   Subsurface stratigraphy

Ground Penetrating Radar

*   Locating buried objects
*   Measuring depth to shallow water table

- Detecting buried containers and leaks
- Delineating solution channels and cavities

Magnetometry

- Locating buried steel containers (55-gallon drums)
- Locating abandoned wells
- Locating pipes and tanks

Metal Detectors

- Locating metallic containers
- Locating buried metallic tanks and pipes
- Locating abandoned wells

Gravity

- Locating buried alluvial sediments in bedrock
- Delineating subsurface cavities
- Estimating recharge and specific yield

4.5.2 GEOPHYSICAL WELL LOGGING

Borehole geophysical survey methods have a number of uses in hydrogeology. For example, they can be used to identify high-permeability zones, clay strata, salinity zones, changes in water quality along a borehole, directions of vertical flow in a borehole, and lithologic correlations. The most commonly used borehole logs in hydrogeologic investigations are caliper, gamma, receptivity, electromagnetic induction, flow meter, televiewer, and neutron. Newer geophysical tools that have great promise in hydrologic studies include nuclear magnetic resonance probes and horizontal flow meters. Use of a combination (often referred to as a *suite*) of geophysical logs is preferred over the use of any single log.

4.6 LOCATING AND TESTING WATER WELLS

In many cases, groundwater for agriculture, industry, and human consumption is extracted from wells. The geological framework, climate, distance to a stream, relation to areas of groundwater recharge and discharge, and topography will affect the best location for a well. Finding the proper location of a water supply well is one of the best ways to ensure a safe and adequate water supply. The well owner's and hydrogeologist's responsibility is to locate the well so as to avoid contaminant sources (unless the purpose of the well is to monitor contaminant levels).

Several techniques can be used to locate a potential water-supply well groundwater source and to determine depth to water, yield, and water quality. The topography may offer clues to the hydrogeologist about the occurrence of shallow groundwater. Conditions for the presence of large quantities of shallow groundwater are more favorable under valleys near rivers and streams than under hillsides or hilltops. In parts of the arid southwestern United States, the presence of "water-loving plants" such as salt cedar and cottonwood trees indicates shallow groundwater. Areas that have saline soils springs, wetlands, and also seeps reflect the presence of groundwater.

- Local geology can offer valuable clues to the potential for obtaining a good supply of groundwater. The hydrogeologist should prepare maps and cross-sections showing the distribution of the rocks both on the surface and underground. As a rule, the wells with the highest yield tap groundwater in alluvial and limestone (karst) aquifers.
- The hydrogeologist obtains information on the wells in the area: their locations, depth to water, well yield, and geology of rocks penetrated by wells. Geophysical well logs are useful to define aquifer characteristics and to correlate water-producing or confining zones.

Wells that fail are typically too shallow, are completed (screened or otherwise terminated in porous media) with small effective porosity or hydraulic conductivity, or do not intersect water-bearing fractures. If usable water is available at a greater depth, then the well can be deepened. Often, wells that are drilled and completed in bedrock cannot be easily deepened, but they sometimes can be made more productive by increasing the fracture-permeability by a technique called hydrofracturing. Well failure may also be due to poor well design or construction, incomplete well development, or blocked well screen. Well failure may also be due to bacteria or mineral encrustation on the screen. This develops after a period of time in some wells.

An experienced water-well driller or hydrogeologist should be consulted to find the best places for drilling wells. Field inspection followed by analysis of geological maps and other available documentation allows the hydrogeologist to recommend where and how to install an efficient water well. In many situations, it may be necessary to drill one or more small-diameter test holes to find the best site for a well that will provide an adequate quantity of good-quality water, especially if a large supply is needed for a town or industry. The test well may be used later for monitoring water-level changes. As a general rule, a hydrogeologist should be involved in the planning and drilling of irrigation, municipal, or industrial wells.

In many places the well location must be approved by local authorities, and well logs and descriptions must be sent to a government agency for entry into their database. This well-drilling database is an available source for hydrogeologists and drillers to locate sites for new wells. Public agencies are a source of information when planning a site for a well. These agencies have compiled data and reports on wells and the groundwater conditions, such as well logs, hydrogeologic maps, basic data, water-level measurements, and interpretive reports (see Chapter 3). In many countries, information on groundwater is available from the national or regional geological survey or water supply authority.

Many countries have well-construction regulations that dictate minimum distances between water wells and potential contamination sources, such as sewer lines, septic tanks, buried fuel tanks, animal waste-storage facilities, waste-treatment ponds, or landfills. Wells should be located as far as possible from potential sources of contamination to protect the water supply. Knowledge of groundwater flow direction and hydrogeology is needed to make an evaluation of possible water-quality degradation.

4.7 WELL DRILLING AND CONSTRUCTION

Large underground tunnels, or *kanats*, were used to collect groundwater in Iran, Egypt, and Persia as early as 800 B.C. (Morgan, 1990). The construction of these infiltration galleries, which collect water from alluvium, was a great achievement. Many of these collection galleries are still in use in Iran and Afghanistan.

The four main types of wells include dug wells, drilled wells, driven wells, and horizontal wells:

1. *Dug wells* are still constructed in some parts of the world. These wells are usually very shallow because they are excavated by hand shovel to below the water table, and digging is ceased when the amount of water entering the hole exceeds the digger's bailing capacity. The well must be lined with rock or cement to keep it open. Because these wells only just penetrate the water table, they often go dry during periods of drought.
2. *Drilled wells* can be constructed in a variety of ways. Water-well drilling (churn drilling) was developed in ancient China. Most modern wells are drilled by cable-tool or rotary-drill equipment. In cable tool drilling, a long, chisel-shaped object is repeatedly dropped into the hole to break up (churn) the rock. In rotary drilling, a bit is mounted to the lower end of the drill rod and rotated to crush and loosen the rock as the hole is deepened. Rotary drilling rigs can drill to depths of more than 900 meters (nearly 3,000 feet). Other types of drilling include reverse-air rotary and downhole hammer.
3. *Driven wells* are constructed by pushing a pipe into shallow sand and gravel aquifers (depths of up to 20 meters, or about 66 feet). A screened well point is attached to the bottom of the pipe before driving. These wells are relatively simple and economical to install, but they can tap only shallow groundwater.
4. *Horizontal wells* are frequently found beneath stream beds in alluvial aquifers. The water produced from these large-volume wells is mostly induced stream flow.

In many countries, drillers subsidized by the government or by international agencies are employed in constructing rural village wells. The most successful drillers use jetting equipment and pumps to construct wells. In some countries, jackhammers and dynamite are used to deepen wells in hard rock. In a city area or large agricultural or industrial area, a hydrogeologist should plan and lead a detailed investigation of the project area.

Other types of drilling include reverse-air rotary. A surface casing is installed usually to a depth of 5 to 15 meters (16 to 50 feet) to protect the drilled well from surface contamination. The annular space between the surface casing and the borehole is sealed by cement grout or clay slurry.

Well casing serves as a support to prevent caving in of the borehole in loose, unconsolidated material or to seal off water of undesirable quality. A well screen in the water-producing zone at the end of the casing (with an appropriate gravel or sand pack) prevents inflow of silt or fine sand while allowing water to enter the well. A sand or gravel pack may also be placed outside the casing and well screen. In fractured-rock aquifers, casing is commonly is not used, with exception of surface casing. Such "open-hole" well construction allows for the maximum number of water-producing fractures to be accessed but can also allow mixing of water from different zones of the aquifer. Such mixing may introduce poor-quality water into a good-quality zone and will result in a water level that is a composite of the potentiometric levels in each of the fractured zones.

4.8   WELL DEVELOPMENT AND TESTING

Any form of well construction causes damage to the aquifer in the vicinity of the well, reducing the aquifer conductivity and well efficiency. For example, if drilling mud is used to keep the borehole open during drilling, the mud can invade the aquifer and clog the pore spaces. This mud must be removed to allow a good connection between the well and the aquifer material.

The process of repairing the aquifer is called *well development* and can be accomplished by several methods. The basic method involves taking such action that alternates the water flow into the well with a reverse flow into the aquifer. Development is an appropriate completion activity for wells that tap sand and gravel aquifers as well as for wells installed in bedrock. Acid treatment and well blasting are sometimes used to develop wells drilled into consolidated rock. The design of a well intended to provide drinking water should include provisions for sanitary protection. Contaminated water from surface runoff from the ground can enter the well through the annulus (opening) between the casing and well bore. To reduce the potential for ground contamination, a cement grout or bentonite seal is placed in the annulus and a cement seal is placed at the surface around the well head.

When the well drilling is completed, the well is bailed or pumped to develop the well and determine the yield. After the screen and sand pack are in place, the well should be surged gently. Many parts of the well need further work after drilling to remove fine material remaining from the drilling so that water can readily enter the well. Possible development methods include compressed air, bailing, jetting, surging, or pumping. The quantity of water being produced by the well is measured during the development, and the procedure is continued until the water is clear of sediment.

After the well development phase is completed, the well is tested to determine how much water it can produce (the well yield). The yield and depth to water during pumping are measured to compute specific capacity—yield is divided by drawdown, or you can use a graphical method of data computation (Moore, 2002). Water samples should be taken at the end of the test and analyzed for chemical, radiological, and biological components.

The space between the borehole and casing (annulus of well) must be sealed to prevent surface contamination from migrating downward and contaminating the water supply. Specific capacity is reported as the discharge rate per unit of draw-down, such as liters per second per meters of drawdown.

4.8   DETERMINING WELL YIELD

Well yield is determined by a specific-capacity test, in which the pumping rate and water-level charges are monitored for a set period of time. The first step is to measure the initial water level in the well. Commonly, a well is pumped at several increasingly greater rates for uniform periods (typically 1 hour) to establish a pumping rate for a long-term aquifer test. The well is then pumped at a steady rate, and the water-level changes are monitored at the pumped well.

4.9   WELL MAINTENANCE

Wells require regular maintenance to perform properly. Maintaining a well involves early detection and correction of problems that could reduce well performance. Records should be kept on well installation, logs, casing and screen location, aquifer tests, and water analyses. A number of factors affect well yield: wear of pump parts, encrusting deposits on the well screen, and corrosion. Periodic chemical analysis of the water can indicate the presence of chemicals that can cause encrustation in wells or could suggest water quality changes in the water due to tapping of different groundwater systems. One method to remove encrustation is to put acid into the well and agitate the water in the well bore. Treating a well with chloride (Clorox) can be an effective way to loosen bacterial growths and slime caused by iron bacteria.

4.10   GROUNDWATER QUALITY

Large amounts of groundwater quality data can be accumulated during a hydrogeologic investigation or a groundwater monitoring program. It is important that the data be organized and checked for technical soundness. Project goals and data evaluation often are dictated by regulatory requirements. Validation of water quality data is critical for the correct assessment of the groundwater quality data (Ohio EPA, 2009). Validation consists of editing, screening, checking, auditing, verification, and certification.

The chemical character of groundwater is an important element in field hydrogeology investigations. The water quality data are used to determine the suitability of the water as a potable (drinking water) supply, for irrigation and commercial use, to establish background conditions (pristine or altered by man), to detect the presence of contamination, and to monitor the effectiveness of remediation of the contamination (Wood, 1976).

When sampling streams and springs, it is important to collect the sample from a freely flowing place as close to the source as possible. Sources of error in sample collection results are from materials and methods used in sampling and from methods used in sample analysis. When sampling a well, the water must be *purged*, or flushed from the well, to get rid of water that has been standing in the well bore.

Ideally one should flush three times the volume of the bore hole. This means pumping the well at least 30 minutes, depending on the diameter of the well and depth to the water table. To obtain a representative water sample from the aquifer, the temperature, pH, and specific conductance should be constant at the time of sample collection. Personal judgment should be used in determining the number of well volumes that should be removed to remove stagnant water. This number of well volumes is controversial. If the sample contains visible undisclosed solids, it should be passed through a pump filter. All samples should be marked immediately with the date and location of collection. Sample bottles should be completely filled with water and contain no air bubbles.

Some properties or constituents in groundwater may change dramatically within a few minutes after sample collection. Immediate analysis in the field may be required if dependable results are to be obtained. Samples may be stabilized by preservative treatment. Some examples of treatment are refrigeration of samples intended for toxic metals such as mercury and addition of acid to prevent the precipitation of metal ions (Wood, 1976).

Groundwater sampling is more complicated than simply removing a volume of water from a monitor well, pouring it into a sample container, and shipping it to an analytical laboratory. Samples must be representative of in situ conditions. Errors may result or contamination may be introduced by the material and methods employed in sampling or in sample analysis. The spatial and temporal heterogeneities in the hydrology and geochemistry of the groundwater environment can also affect the chemical character of the water. Materials for well construction and sampling equipment should be selected so as not to interact or interfere with the constituents being analyzed. The recommended rigid materials in sampling applications are Teflon (flush threaded), stainless steel (flush threaded), and PVC (flush threaded).

Low-flow sampling from monitor wells is a new technique based on the use of a submerged pump that can be adjusted at rates from 100 ml per minute to 1 ml per minute. The objective of this technique is the recovery of representative samples of the aquifer adjacent to the well screen. Ideally the flow rate of water from the pump will approximate the water entering the well from the aquifer.

4.11   FIELD MEASUREMENTS OF WATER QUALITY

The tests for groundwater quality made in the field are temperature, dissolved oxygen, pH, Eh, alkalinity, and specific conductance. With portable gas chromatograph instruments, many constituents can be analyzed at the well site.

Temperature is important for many reasons. For example, temperature measurements are critical to indicate inflow of surface water into the well.

Dissolved oxygen (DO) is an important parameter for surface water and groundwater. Aquatic fauna need DO to survive, and generally, the higher the reading the better the quality. Measurements of DO should be made with a field kit immediately after the sample is taken.

A *pH* meter must be carefully calibrated with standard buffers of pH 4, 7, and 10 before a measurement is taken. Failure to calibrate will result in meaningless data.

Eh is also known as Redox potential and is a qualitative measurement of reduction oxidizing conditions in a groundwater sample, which in turn is an indicator of the amount of oxygen kept in solution.

4.11.1 SPECIFIC CONDUCTANCE

Specific conductance is the numerical expression of the ability of the water to conduct an electrical current. Conductance measures the total dissolved ion concentration in water. Field determinations of specific conductance can be an aid in choosing additional sampling sites and frequency of sampling. Measurements can indicate if sufficient water has been pumped before a sample is collected and that the chemical and physical characteristics of the quality of the water in the well have been stabilized. Conductance meters should be battery operated, equipped with a temperature compensator, and read in micromohs at 25°C. The conversion factor from conductivity to total dissolved solids TDS conversion factor 0.5 and 0.75.

When sampling streams and springs, it is important to collect the sample from a freely flowing stream or as close to the source as possible. When sampling a stream, hold the open end of the bottle downstream and cap it underwater to help eliminate air bubbles. When sampling in a lake, invert the bottle and push it down under the water surface. When the whole bottle is submerged, turn it upright and allow it to fill. Measure all related field parameters at the same time you collect the sample.

4.12 TESTING THE QUALITY OF GROUNDWATER

Water quality is commonly characterized as *safe* or *good*. *Safe water* typically means that the water is free from bacteria and disease-causing organisms, as well as from minerals and substances that can have adverse health effects. The term *good-quality water* is a relative term whose meaning depends on the intended use of the water. Sanitary quality can be assessed by periodically analyzing water samples for coliform bacteria and nitrate. These substances do not normally occur in groundwater, and their presence may indicate contamination.

Coliform bacteria are useful indicators of harmful microorganisms. According to drinking-water standards of the United States, safe drinking water should not contain more than one coliform bacterium per 100 ml (6 cu in.) of water. If coliform bacteria occur in well water in numbers that indicate the water is unsafe, the water should be disinfected either by chlorinating or by boiling before drinking until further sample analyses indicate that the water is safe to drink.

High nitrate levels may indicate organic contamination from nearby sources of nitrate, such as barnyard drainage, animal waste storage, percolation from agricultural land, fertilizers, and septic tanks. The U.S. national limit for nitrate–nitrogen concentration is 10 mg/l, which is equivalent to 45 mg/l of nitrate (Fetter, 1994). The European standard is 11.4 mg/l nitrate–nitrogen (50 mg/l as nitrate). The main reason for this limit is because of the risk to (young) infants ("blue baby" disease) from drinking high-nitrate waters.

Well water can be analyzed either by a county or state public health environmental laboratory or by a qualified private laboratory. Municipal authorities (local health departments) have information on where water may be analyzed. Routine tests for coliform bacteria or inorganic contaminants, such as nitrate and salts, are relatively inexpensive. However, analyses for inorganic and organic industrial chemicals and pesticides can be very expensive. Those concerned about possible contaminants in their water supply can contact their local health departments for advice on having their water tested for safety. Specialists from these departments can help determine if a cause for concern exists before extensive and potentially expensive water testing is performed. Keeping accurate records of the water tests ensures proper documentation of the water-quality history of a particular well. Even without obvious signs of contamination it is advisable that owners or users of water from wells that are not regularly tested have their water analyzed at least once a year.

If water contamination persists even after several testings, an alternative drinking water supply should be used or a new source obtained. Leading causes of bacterial and nitrate contamination are improper well construction and/or poor well location. In most cases, nitrate contamination is limited to the shallow part of the water-bearing formation. When the contaminant is a volatile organic compound (VOC), it may cause problems, not only by drinking but also by inhalation and absorption through the skin.

ADDITIONAL RESOURCES

ASTM, 1993, D 5474 *Standard guide for selection of data elements for groundwater investigations.*

ASTM, 1995, D 5777 *Standard guide for using the seismic refraction method for subsurface investigation.*

ASTM, 1996, D 5979 *Guide to conceptualization and characterization of ground-water flow systems.*

Driscoll, F. G., 1986, *Groundwater and wells.* Johnson Division, 1108 pp.

Duttro, J. T. et al., 1989, *AGI data sheets for geology in the field, laboratory and office,* American Geological Institute.

EPA, 1993, *Subsurface characterization and monitoring techniques, Vol. 1: solids and ground water, appendices A and B,* Cincinnati, OH: Center for Environmental Research Information, EPA/625/R-93/003a.

Fetter, C. W., 2001, *Applied hydrogeology,* fourth edition. Upper Saddle River Cliffs, NJ: Prentice Hall, 598 pp.

Haeni, F. P., 1988, Application of seismic-refraction techniques to hydrologic studies, in *USGS Techniques of Water-Resources Investigations,* Book 2, Chapter D2, 86 pp.

Keys, W. S., 1990, Borehole geophysics applied to ground-water investigations, in *USGS Techniques of Water-Resources Investigations,* Book 2, Chapter E2, 150 pp.

Keys, W. S., and MacCary, L. M., 1971, Application of borehole geophysics to water resources investigations, in *USGS Techniques of Water-Resources Investigations,* Chapter E1, 126 pp.

Mandel, S., and Shiftan, Z. L., 1981, *Groundwater resources.* Academic Press, 269 pp.

McGuinness, E. L., 1969, *Scientific or rule-of-thumb techniques of ground-water management: which will prevail?* U.S. Geological Survey Circular 608, 8 pp.

Ohio EPA, 2009, Monitoring well development, maintenance, and redevelopment, in *Technical guidance manual for groundwater investigations,* Chapter 8.

Sanders, L. L., 1998, *Field hydrogeology.* Prentice Hall, 381 pp.

Sara, M., 1989, *Site assessment manual,* Waste Management of North America Inc.

Summers, P., 1985, *Guidelines for conducting ground-water studies in support of resource program activities*, U.S. Department of Interior, Bureau of Land Management.

U.S. EPA, 1978, *Electrical resistivity evaluations at solid water disposal facilities*, USEPA Report SW-729.

U.S. EPA, 1991, *Site characterization for subsurface remediation*, Office of Research and Development, EPA/625/4-91/026. (1)

U.S. EPA, 1993, *Subsurface characterization and monitoring techniques: a desk reference guide, Volume I, Solids and ground water, appendices A and B.*

U.S. EPA, 1993, *Subsurface characterization and monitoring techniques: a desk reference guide, Volume II, The vadose zone, field screening and analytical methods, appendices C and D.* (3)

Weight, W. D., 2008, *Manual of field hydrogeology*, second edition, McGraw Hill.

Weight, W. D., and Sonderegger, J. L. 2001, *Manual of applied field hydrogeology*, McGraw Hill, 608 pp.

Zody, A., Eaton, G., and Mabey, D. R., 1974, Application of surface geophysics to ground water investigations, in *USGS Techniques of Water-Resources Investigations,* TWRI Book 2, Chapter D1, 116 pp.

5 Subsurface Investigations

The fieldwork for a hydrogeologic investigation may be divided into several phases: defining project objectives, geologic mapping, inventory of wells, logging of wells, observations of water levels, collecting water samples for chemical analysis, collecting data on amount of pumpage and use of water, test drilling, and aquifer testing. Some of these phases may be undertaken simultaneously, for example, inventorying of wells, measurement of water levels, and collection of water samples. Some phases may be interdependent. The amount of test drilling needed will depend on the number and locations of existing wells and the adequacy of information on them.

5.1 GEOLOGIC MAPPING

The geological framework is the key to any groundwater investigation. It follows that geologic mapping, of both the surface and subsurface (cross-sections), is one of the first field phases of an investigation. In most areas of the United States, some geologic mapping has been done and serves as a basis for that required in a groundwater study. Particular attention is given to the geologic units that will most affect the occurrence, movement, and quality of groundwater.

At the start of many investigations, subsurface mapping of aquifer thickness and configuration is often lacking. The subsurface is mapped in cross-sections from logs of wells with the aid of geophysical survey techniques. The logs collected from well drillers, well owners, public-agency files, and oil and gas exploration companies are useful. If possible, wells for which no logs are available should be logged by geophysical equipment and correlated with known geology in the vicinity of the well.

5.2 INVENTORY OF WELLS

A well inventory is a detailed list of all known wells, test holes, and springs in the project area (Figure 5.1). The list is made by contacting well owners, well drillers, the state engineer, and the U.S. Geological Survey (USGS) and other federal agencies. It includes those wells in use and abandoned wells. The inventory for each well should include the following: owner's name, address of the site, map reference, well construction information, water level, drilling method, diameter, depth, screen, aquifer description, lithological and structural characteristics, geology, water quality, yield, GPS coordinates, and annual withdrawal. Each well entry should include a unique number. In some cases this would be the latitude and longitude. Geographic positioning systems (GPS) should be used whenever possible as an aid to determine location and altitude. It should be noted, however, that although GPS locations are very accurate, altitude obtained from most hand-held GPS units can be in error by several meters (tens of feet).

Date _____Field number _____

Recorded by _____

Source of data_____

USGS quad sheet location: State _____ County _____

_____ 1/4 _____ 1/4 SEC _____ T _____ R _____

Owner _____Address _____

Tenant _____Address _____

Driller _____

Topography _____Elevation _____

Drilling method _____Depth _____

Casing: Diameter _____Inch type _____

Depth to _____Finish_____

Aquifer _____

Water level _____ Date _____ _____ Below/above surface

Pump _____ Power _____Yield _____

Use of water _____Adequacy _____

Quality _____Temp _____

Remarks _____

FIGURE 5.1 Well inventory form. (From Moore, J., *Field Hydrology: A Guide for Site Investigations and Report Preparation*, CRC Press, Boca Raton, 2002. With permission from Taylor & Francis.)

While the geology is being mapped, the existing wells and springs in the area are scheduled (examined); that is, the area is visited and all information is recorded relative to elevation, depth, diameter, location, age, water level, pumping rate and drawdown, use, and formations penetrated. Water samples are collected from wells and springs representative of various aquifers. These samples are sent to a laboratory for analysis. If a water quality problem is known or suspected, partial analyses may be made in the field to determine the best locations for more extensive and detailed sampling (Figure 5.2). In addition to the water-level measurements made at the time wells are inventoried, water levels in selected wells are measured periodically, and automatic recorders are installed on some to determine magnitudes of fluctuations. Also, owners of irrigation wells are canvassed to determine the amount and rate of withdrawal of water from the area. The well owner can be a good source of information to locate existing and abandoned wells in the area.

5.3 MONITOR WELLS

Monitor wells are designed to obtain representative groundwater information and water quality samples from specific aquifers at selected depth intervals. Monitor wells constructed and developed following standard practice should produce relatively turbidity-free samples (Figure 5.3). A suggested monitor well design is given in ASTM (1990, p. 952), and a detailed discussion of monitor well construction is given in Nielsen (1991).

- Team members

- Purpose of sampling

- Location (topographic map)

- Type of site

- Name and address of field contact

- Date and time of sample collection

- Weather conditions

- Purging information

- Sampling device

- Sample containers

- Sample treatment

- Photograph

FIGURE 5.2 Items in field notebook. (From Moore, J., *Field Hydrology: A Guide for Site Investigations and Report Preparation*, CRC Press, Boca Raton, 2002. With permission from Taylor & Francis.)

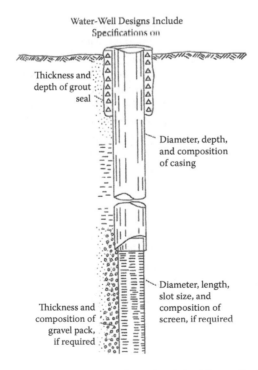

FIGURE 5.3 Well construction. (Redrawn from the United States Geological Survey.)

The objective is to construct a well for which you can have confidence in water-level and water quality data and samples and that truly represents the zone of interest. The objective for the monitor well (or monitor well program) must be well defined to be achieved. Some questions that can be answered from of the information from a monitoring project are as follows:

- How many aquifers are penetrated by the well?
- Where is the contamination?
- Are there perched zones (saturated lenes or layers of deposits above the local water table)?
- If any contaminants are present, at what depth is the contaminant, and what is the nature of the contaminant?
- On the basis of water-level measurements and gradients, where are the recharge and discharge areas for groundwater in the study area?

5.4   TEST DRILLING AND EXAMINATION OF DRILL CUTTINGS

If the data from existing wells are so incomplete that interpolation and interpretation cannot fill the gaps, it may be necessary to drill new exploratory holes. Information and data collected during the drilling of new wells will make possible the definition of the lateral extent of or change in subsurface formations, the collection of rock and water samples otherwise unobtainable, and the opportunity to conduct pumping (or aquifer) tests. In alluvial-filled valleys, it is generally desirable to drill a series of test wells perpendicular to the axis of the valley. Preliminary surface-geophysical surveys should be used to help determine the location of test wells.

Detailed information on the hydrogeology for an area can only be obtained by drilling test holes and wells. The value of the information depends on the care exercised in collecting and examining the cuttings and the samples, and the accuracy and completeness of the description of the sample. The more subsurface information gathered for the area, the better the understanding of the geological framework and the greater the success in locating groundwater supplies. It is helpful to be familiar with different drilling methods before attempting to log your first test hole. You will need to know how the cuttings are moved to the surface. For example, as drilling proceeds, the cuttings near the drill bit are mixed with cuttings from earlier drilling. This yields a mixture of cuttings that you must interpret. A hand lens, and in some cases a microscope, should be used to examine the cuttings. A field geotechnical gauge should be used to describe the color and grain size (sieve size) for consistency. When preparing the lithologic log, the rock type is recorded first (sandstone, shale, limestone, etc). The color of the rock is then listed. The Wentworth grade scale is used to determine the size of particles.

The following information should be shown on the log of a test well:

1. Driller, agency, or consultant
2. Location and number of the well

3. Starting and finishing date of well drilling and completion
4. Depth of well
5. Depth to water
6. Results of pumping or bailing tests
7. Temperature and water-quality field data
8. Depth to "first water"
9. Depth of water table as drilling is in progress
10. Fluid density, viscosity, temperature, and specific conductance

The importance of recording all such data before they are lost cannot be emphasized too strongly. Another objective of the test drilling is to identify when water is first observed and where it might be coming from. The hydrogeologist should measure water level in the test hole at the start of each day.

5.5   WATER-LEVEL MEASUREMENTS

Water-level measurements in observation wells are the principal source for information about the hydrologic stresses acting on aquifers and how these stresses affect groundwater recharge, storage, and discharge. Long-term systematic measurements provide essential data needed to evaluate changes in groundwater over time, to develop groundwater models and forecast trends, and to design and monitor the effectiveness of groundwater management and protection programs.

Following are some uses of water-level data for groundwater:

1. Indicate the change in groundwater storage
2. Observe the rate of regional groundwater withdrawal
3. Show relationship of groundwater and surface water
4. Provide long-term records used to evaluate effect of water management program
5. Estimate recharge and discharge
6. Estimate the hydraulic character of aquifer
7. Provide database for water management needs
8. Identify areas where water level is near land surface
9. Estimate rate and direction of groundwater movement
10. Delineate reaches of gaining and losing streams
11. Monitor earthquakes

The number and location of observation wells are critical to any water-level data program. In selecting the location and depth of observation wells, the physical boundaries and geologic complexity of aquifers in the area should be considered. Multilayer aquifer systems may require measurements in wells completed at multiple depths in different hydrogeologic units. Commonly overlooked is the need to collect other types of hydrologic information as part of a monitoring program (such as rainfall and stream flow). Observation wells should be selected with an emphasis on wells for which measurements can be made for an indefinite time.

Good quality-assurance practices help to maintain the accuracy and precision of water-level measurements and ensure that the wells reflect conditions in the aquifer being monitored and provide data that can be relied on for many intended uses. Field practices that will ensure quality of data include the establishment of permanent reference points, periodic inspection of the well structure, and periodic hydraulic testing of the well to ensure connection with the aquifer.

Manual measurements of water levels in observation wells can be made by one of the following methods:

1. Wetted steel tape
2. Air-line submergence
3. Electrical Tape
4. Pressure transducer

5.5.1 Wetted Steel Tape

Before the 1960s, most water-level measurements were made with a steel tape (most likely a 30 or 60 meter Lufkin tape). The tapes have been replaced by electrical methods and pressure transducers. However, the steel tape still has applications today, for example, to calibrate pressure transducers. Before the tape is lowered down the well, the lower 0.30 to 0.60 meters or so of the tape is coated with carpenter's chalk. The tape is lowered down the well until the lower part is submerged. The tape is held on at the measuring point (MP), and this value is noted. The tape is pulled back to the surface, and the wet mark on the tape (e.g., 1 meter) is recorded. The MP is usually the top of the casing, and the distance of the MP from the land surface is noted on the well measurement form. If the MP is changed, a note to that effect should be noted on the measurement form. The water measurement should be repeated at least three times to ensure a precise depth to water value. Measurements are made to 1 mm accuracy. Some problems that may be encountered using this method are moisture on the well casing, cascading water, and oil (leaking from pump lubrication). For measurements of waters levels made at depths of greater than 60 meters feet should refer to Garber and Koopman (1968).

5.5.2 Electrical Tape

Most electrical tapes are marked every hundredth of a foot or meter. An electrical probe is lowered into the water, which completes an electrical circuit and causes either a buzzer to sound or a light to illuminate (Figure 5.4). If electrical tapes are used in deep wells (greater than 70 meters), stretching of the wires could result in small errors in the measurement.

FIGURE 5.4 Observation well measurement.

5.5.3 PRESSURE TRANSDUCER

The transducer probe enables measurement of depth to water in a well by sensing the weight of the column of water above the probe. Transistors are very useful in measuring water levels during aquifer tests when the water-level changes rapidly with time. Data are transferred to data loggers like the In-Situ logger (Figure 5.5). A wide variety of pressure transducers are available, and care should be taken in choosing the transducer needed for a particular project. Freeman and others (2004) describe the use of submersible pressure transducers in hydrologic investigations.

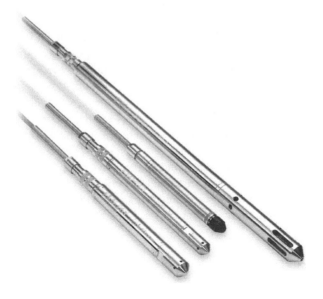

FIGURE 5.5 Pressure transducers. (Courtesy of In-Situ, Inc.)

5.6 TRACING TECHNIQUES

5.6.1 INTRODUCTION

In recent years groundwater tracing techniques have been used in a variety of hydro-geologic settings to aid in characterizing groundwater flow systems. Tracing techniques have been demonstrated to be especially useful in karst and fractured rock hydrogeologic settings. Tracer recovery data are useful for identifying and characterizing groundwater flow paths, flow velocities, boundary conditions, time since recharge, and groundwater discharge to streams and other surface waters. During the past 3 decades, tracer tests have been increasingly used to investigate solute transport processes in groundwater, including multispecies reactive transport, colloid-facilitated transport, and pore scale mixing (Devine & McDonnell, 2005). Tracer test results can also be used to refine hydrologic models of groundwater flow. Groundwater tracing techniques measure properties in situ, generate empirical data, and make fewer assumptions about hydrogeologic conditions than do hypothetical (e.g., Darcy) or numerical simulations. Tracer recovery data, when combined with discharge data, can provide quantitative data that can be useful for assessing the relative importance of various flow paths and sources of recharge. Tracer tests can also be used effectively to design or test a groundwater monitoring system. Tracer test results depend on the conservative nature of the tracer, unambiguous detection, proper test design, and holistic interpretation. In designing a groundwater tracing test, it is important to consider the types of injection and recovery sites. Tracers can be introduced and sampled for at natural sites—springs, streams, karst features, and so forth—or artificial sites such as wells, trenches, mine portals/shafts, and water supply facilities. Tracer test design should also consider whether to rely on the natural gradient to move the tracer or to create a forced gradient by raising or lowering the water table.

In general, tracing can be divided into two categories: label tracing and pulse tracing. Label tracing allows for the identification of specific groundwater or contaminant plumes based on a specific tracer that occurs in the groundwater. Pulse tracing involves introducing a tracer into a groundwater flow system at concentrations significantly above background. Groundwater tracers can be divided into two types—natural and artificial. Natural tracers are more applicable for labeling waters, whereas artificial tracers are more suitable for pulse tracing. Important natural tracers include stable and radioactive isotopes, selected ions in solution, selected field parameters (specific conductance and temperature), and selected microorganisms. Commonly used artificial tracers include organic dyes and dye intermediates, fluorocarbons, gases, and salts (chloride, bromide, etc.). Several tracers can be used together, allowing several potential pathways to be evaluated simultaneously.

5.6.2 NATURAL TRACERS

"Environmental" isotopes. Clark and Fritz (1997) refer to "environmental" isotopes as the naturally occurring isotopes of elements found in abundance in the environment. Stable isotopes of hydrogen, oxygen, nitrogen, sulfur, and carbon have proven

to be versatile natural tracers. For a given element, the ratio of the two most abundant isotopes can vary due to partitioning or fractionation related to differences in isotopic mass and reaction rates. This allows the ratios of isotopes of the same element to become fingerprints of climatic and hydrological conditions. Stable isotopes of oxygen and hydrogen (18O / 16O and 2H / 1H) are ideal tracers of water sources and movement because they are integral constituents of the water molecule. Stable isotopes of oxygen and hydrogen behave conservatively because interactions with organic and geologic material along the flow path will have a negligible effect on the ratios of isotopes in the water molecule. Isotope ratios for 18O / 16O and 2H /1H are expressed as the difference (δ) between the measured ratio of the sample and reference over the measured ratio of the reference. Because fractionation processes do not cause great variations in isotopic ratios, δ values are expressed as parts per thousand or per mil (0/00) difference from the reference. Variations in altitude, latitude, and annual rainfall/snow melt strongly affect the 18O / 16O composition of recharge waters. Isotopic fractionation can be more pronounced in low-temperature settings than in warmer settings. Ratios of 18O / 16O can also be measured for oxygen in molecules other than water, including commonly dissolved constituents such as sulfate (SO_4) carbonate (CO_3) and nitrate (NO_3).

Stable isotope ratios for 34S / 32S, 15N / 14N, and 13C / 12C are also useful tracers. These isotopes are constituents that are dissolved in water or carried in the gas phase. The ratios of stable isotopes of dissolved sulfur, nitrogen, and carbon can be significantly altered by reactions with organic and geologic material; however, data for these isotopes can provide information on the reactions that are responsible for their presence in the water and the flow paths implied by their presence. Isotopic ratios for these elements are also expressed as the difference between the sample and a reference. The 13C / 12C ratio can be useful for tracing carbon sources and reactions for interacting organic and inorganic species. Ratios of 15N / 14N are useful for evaluating nitrogen reactions in water. For example, the relationship between the 15N / 14N ratio and the 18O / 16O ratio of nitrate can be used to help distinguish sources of nitrate, and the 18O / 16O ratio of nitrate is a good tracer of gentrification. The ratio of 34S / 32S in dissolved sulfate, sulfide (HS–) and hydrogen sulfide gas (H_2S) can be used to help evaluate landfill leachates and acid mine drainage and dating sulfate-reducing groundwater.

For detailed information on isotope geochemistry, the processes affecting isotopic composition, and applications of isotope hydrology, refer to Kendall and McDonnell (1998) and Clark and Fritz (1997).

Radioactive isotopes. Tritium (3H), a radioactive isotope of hydrogen (half life = 12.43 years), is naturally produced by cosmic-ray spallation in the upper atmosphere (bombardment of nitrogen by neutrons). Several hundred kilograms of tritium were also released to the atmosphere during the period of above-ground thermonuclear testing, which was at its peak during the 1950s and was discontinued in 1963. About 80 kilograms of this isotope are still present on the earth's surface at this time. As a result, the tritium input function is a large spike, with highest concentrations in the early 1960s. Tritium is the only conservative tracer that can be used to determine the age of water older than 1 year. Since it is part of the water molecule, it is the only direct water dating method. Prebomb concentrations of tritium in water have been determined to be close to the detection limit of 1 tritium units (TU). The presence of tritium above background levels in groundwater is an indication that recharge occurred during or after the bomb-testing period. Tritium concentrations in water are expressed as absolute concentrations, using tritium units. One tritium unit is equal to one 3H atom per 1,018 atoms of hydrogen. Present-day tritium concentrations in precipitation are about 12–15 TU.

In has been over 40 years since above-ground nuclear testing was discontinued, and atmospheric input is now approaching prebomb levels. In addition, radioactive decay of tritium and attenuation by matrix diffusion and dilution by mixing with younger water with less tritium or older water with no tritium make it impossible to obtain an "absolute" age of groundwater. Clark and Fritz (1997) provide a qualitative interpretation of groundwater mean residence times (Table 5.1).

Carbon 14. With a half life of 5,730 years, 14C is a useful tracer for aiding in estimating the age of groundwaters that were recharged in the late Quaternary or later. 14C is produced in the upper atmosphere and enters groundwater as dissolved $14CO_2$. Atmospheric $14CO_2$ mixes with biomass, meteoric waters, and oceans. The primary source of 14C in groundwater is the huge reservoir of 14C in the soil zone, where it accumulates as a result of decay of vegetation and root respiration. Any carbon compound derived from atmospheric CO_2 since the late Pleistocene can potentially be dated. 14C activity is determined for dissolved inorganic and organic carbon—not on the water molecule itself. Groundwater age is based on measuring the loss of parent 14C by radioactive decay and assumes that the initial concentration of the parent 14C is known and that radioactive decay is the only process that changes the concentration of the parent 14C. Precise measurement of initial 14C activity in tree rings and corals has established the initial concentration of parent 14C over the past

---

TABLE 5.1

Qualitative Interpretation of Mean Groundwater Residence Time—Based on Tritium Concentrations

Tritium Concentration (TU) Qualitative Age

Continental regions

<0.8 Submodern—recharged < 1952

0.8–4 Mixture between submodern and recent

5–15 Modern < 5 to 10 years

15–30 Some bomb tritium present

>30 Considerable component of recharge from 1960s or 1970s

>50 Dominantly 1960s recharge

Coastal/low-latitude regions

<0.8 Submodern—recharged < 1952

0.8–2 Mixture between submodern and recent

2–8 Modern < 5 to 10 years

10–20 Residual bomb tritium present

>20 Considerable component of recharge from 1960s or 1970s

Source: Clark and Fritz (1997).

30,000 years. Accurate age dating is constrained by poor preservation of old carbon compounds and subsequent contamination. In addition, the reaction and evolution of carbonate systems can dilute the initial 14C activity (Clark and Fritz, 1997). Anthropogenic activities during the past century have also resulted in further dilution of initial 14C activity. These problems are not simple and result in a complicated analytical methodology that requires corrections to account for the dilution.

5.6.3 ARTIFICIAL TRACERS

Artificial tracers are substances introduced into the groundwater flow system, either purposely as part of a designed tracer test or inadvertently as a spill or other anthropogenic activities. Commonly used artificial tracers include organic dyes, salts (chlorides, bromides), and fluorocarbons. To serve as a suitable tracer a substance must be (1) nontoxic to people and the ecosystem, (2) either not present naturally in the groundwater system or present at very low, near-constant levels, (3) soluble in water with the resultant solution having nearly the same density as water, (4) conservative, (5) easy to introduce into the flow system, (6) unambiguously detectable in very low concentrations, and (7) affordable and readily available. Organic fluorescent dyes are the most commonly used groundwater tracer.

Fluorescent dyes. Fluorescent dyes constitute some of the most analytically sensitive, versatile, nontoxic, and inexpensive artificial groundwater tracers. The use of a fluorescent dye to trace groundwater dates back to at least 1877, when fluorescein was used as a tracer to prove the connection between the Danube River and the Aach spring (Flury and Wai, 2003). Commonly used fluorescent dyes in groundwater investigations include sulpho-rhodamine B, rhodamine B, fluorescein/uranine, rhodomine WT, eosin, and phloxine. Unfortunately, a single dye may have more than one commercial name, and several different dyes may have the same commercial name. The most complete reference for fluorescent organic dyes is *The Color Index* (Society of Dyers and Colorists, 1971, 2009). Dyes are listed and described by color and by chemistry using both the color index (C.I.) constitution number and the C.I. generic name. This classification allows for an unequivocal identification of each dye. It is important to report the C.I. name and number in all scientific reports. Commercial grade, organic, fluorescent dyes can be purchased as a liquid compound or as a powder. Uranine, eosin, and phloxine are approved by the U.S. Food and Drug Administration. Many fluorescent dyes have been approved for use as tracers in aquifers and streams that are used to obtain drinking water supplies. However, some states have may restrict the use of artificial tracers in groundwater.

Fluorescent dyes are detectable at very low concentrations. Most fluorescent dyes will work well in waters where pH is near neutral. In acidic conditions the fluorescence of some dyes is minimized. However, the dyes will fluoresce again if the pH of the sample is adjusted to more alkaline conditions (Gareth Davies, personal communication). Fluorescent dyes are organic compounds that are composed of large molecules and will interact to some degree with solid soil and sediment surfaces.

Salts. Chloride and bromide solutions are commonly used for both groundwater and surface-water tracing. These salts are very soluble, relatively inexpensive, conservative, and nontoxic at concentrations typically used for tracing, and they are also easily detectable at low concentrations. Chloride occurs naturally in some groundwaters, often in the tens to hundreds of parts per million. Natural concentrations of bromide are much lower. Chloride has commonly been used as a label tracer for contaminant plumes that originate at landfills or other industrial facilities. Stream-tracing techniques, which include the continuous injection of a constant concentration of a tracer (typically bromide), provide very accurate stream discharge measurements, based on dilution of the tracer, and provide detailed information on groundwater inflow zones to streams.

Chlorofluorocarbons. Since the 1930s, chlorofluorocarbons (CFCs), synthetic organic compounds used in numerous industrial and refrigerant applications, have been released into the atmosphere. These compounds, often referred to collectively as freon, dissolve in precipitation and are then distributed throughout the hydrologic system, including groundwater recharge. The most commonly used CFCs for groundwater tracing are CFC 11, CFC 12, and CFC 113. These compounds are nontoxic, nonflammable, and noncarcinogenic. Concentration data for these CFCs can be used to trace the flow of young groundwater by establishing a residence time (time since recharge) or an "apparent" age. The age is established by comparing measured CFC concentrations in groundwater to known historical atmospheric concentrations and/or to calculated concentrations expected in water in equilibrium with air. The accuracy of the age estimate depends, in part, on the degree of attenuation of the CFC compounds along the groundwater flow path. Chemical (microbial degradation) and physical (sorption) processes can significantly control the concentrations of CFCs in groundwater. Because of this, it is important to also collect data on additional constituents such as dissolved oxygen, dissolved methane, dissolved nitrogen, and dissolved argon.

Even though CFCs are nontoxic, they do contribute to ozone depletion. Because of this, production stopped in 1996 as a requirement of the Clean Air Act. The USGS estimates the atmospheric lifetimes of CFC 11, CFC 12, and CFC 113 to be 45, 87, and 100 years, respectively.

5.6.4 FIELD METHODS

The usefulness and appropriateness of a groundwater tracer test depend on the questions to be addressed by the hydrogeologic investigation. Tracer tests are appropriate when groundwater flow velocities are such that results will be obtained within a reasonable period of time—usually less than a year. The usefulness of tracer test results is highly dependent on proper test design (particularly determination of injection and sampling locations), the nature of the tracer, the ability to detect the tracer at low concentrations, and correct interpretation of recovery data. Prior to conducting a tracer test, it is very important to use other geologic and hydrologic data and information to develop a basic understanding of the hydrogeologic setting and the groundwater flow system to be traced. This understanding can then be used to determine (1) the appropriate type of tracer, (2) the tracer injection location and method, (3) appropriate sample collection locations, and/or (4) which stable or radioactive isotopes should be included in the sampling plan. It is always advisable to sample more locations rather than fewer locations. It is also important to know precisely how much tracer mass is injected. This will allow for a determination of the percent of tracer mass recovered at a given sampling location if the flow can be measured at the sampling location. This quantitative aspect of tracing can be important in helping to evaluate the significance of any given groundwater flow path.

Isotopes. Chapter 10 of Clark and Fritz (1997) includes an excellent discussion and comparison of sampling and analytical protocols and procedures for collection of water samples for isotopic analysis. Sample size, filtering, preservation, container type, holding times, and method of analysis vary quite a bit between different stable and radioactive isotopes. In general isotopes of water (oxygen 18, deuterium, tritium) have simpler sampling protocols than isotopes of dissolved inorganic and organic carbon (carbon 13 and carbon 14), dissolved sulphate and sulphide (sulfur 34), dissolved gases (helium, argon 39, krypton 85), and dissolved uranium (uranium 234 and uranium 238). It is very important to note that a given location should be sampled three to four times during an annual hydrograph to use stable water isotope data and tritium data for evaluating flow systems.

Sample collection and analysis, 18O and 2H. Stable water isotope samples are collected in 30 ml borosilicate vials with airtight caps. The δ 18O values are obtained using a CO_2/H_2O equilibration technique on a SIRA Series II mass spectrometer with a nominal precision of 0.1 percent per mil following the protocol of Epstein and Mayeda (1953). The 18O values are expressed in the conventional delta (δ) notation as the per mil (0/00) difference relative to the international Vienna Standard Mean Ocean Water (VSMOW) standard: d (0/00) (Rx /Rs 1) × 1,000 where R denotes the ratio of heavy to light isotope in a sample Rx and the standard Rs. The 1 s precision is 0.05 0/00, and accuracy of analysis based on replicate samples is 0.09 0/00.

Sample collection and analysis, tritium. One-liter samples are collected in high-density polyethylene bottles. The bottles are sealed and then analyzed for tritium using liquid scintillation counting. Distilled sample water is reduced electrolytically in Ostlund style electrolysis cells from an initial volume of 200 ml to about 10 ml in a cooling bath. The tritium is retained preferentially to hydrogen in this system, resulting in an increase of the tritium concentration by a factor of 16. The remaining water is mixed with a scintillation cocktail and counted in Packard scintillation counters of the CA 2000 series. The detection limit at the 1 sigma level is about 0.3 TU with an uncertainty of about 3 percent of the sample concentration.

Organic dyes. The use of organic dyes as hydrogeologic tracers requires specific field sampling and analytical procedures. Careful thought should be given to the selection of the proper dye and the method used to introduce the dye into the groundwater. Dye selection is based, in part, on the chemical and toxicity characteristics of the dye. Also, fluorescence is reduced in some dyes when dissolved in low-pH waters, and some dyes will fluoresce better in cooler waters. All of these factors should be considered when choosing an organic dye for use as a tracer.

It is also very important to carefully consider the best way to introduce the dye tracer into the groundwater system. Common methods include (1) injection into a previously constructed or a new well, making sure the well will take water prior to introducing a dye tracer, which can be determined by conducting a simple aquifer test using the proposed injection well; (2) injection into a stream, making sure the stream gradient is low and the reach of stream below the injection point is not a losing reach; (3) injection into a constructed trench, making sure the trench takes water; (4) injection into a sinkhole, in karst terrain; and (5) injection into a mine shaft.

Sample collection protocol for water samples that may contain the dye tracer are relatively simple. Sample containers and storage should minimize all exposure to light to prevent degradation of the dye. Samples do not require filtering or preservation, but water samples that may contain dye should be kept cool until analysis is complete. They should be analyzed within 2 weeks to minimize bacterial degradation of the dye.

Collection of water samples can be done using an autosampler (useful for collecting many samples in a short time and from locations with difficult access) or by grab sampling. Once the water samples are collected, they should be analyzed on a spectrofluorometer to confirm the nature of the fluorescence and then analyzed with wet chemistry methods for dye concentration. It is important to conduct both types of analysis to better confirm the presence of the dye that was injected. Sophisticated sampling techniques can be achieved by using flow-through fluorometers, which measure total fluorescence in water on a real-time basis. This type of sampling requires a power source and data loggers. Portable, dye-specific, submersible fluorometers are also available for installation in wells. The cost of a fluorometer varies from $2,000 to $5,000 for a submersible fluorometer to $20,000 for a spectrofluorometer. Analytical costs for water samples typically range from $15 to $20 for a single dye to $35 to $40 per sample for multiple dyes.

Small bags of activated charcoal can also be used to detect dyes. Organic dyes will sorb onto charcoal if water that contains dye comes into contact with the charcoal. Charcoal bags are placed in water at sampling locations and then retrieved for analysis at selected time intervals. It is important to note that determining the travel time from an injection location to a given charcoal bag is constrained by the time interval between retrieval of the bags.

The most important rule of thumb for sampling is to collect samples often at many places. Collecting samples for organic dye analysis is relatively easy and inexpensive. Since it is not always possible to predict all locations where dye may be recovered, it is best to have more rather than fewer sampling locations. By collecting samples frequently, at important locations, the data can be used to construct breakthrough curves of recovery versus time. Detailed breakthrough curves can be used to do rigorous analyses of the recovery data.

ADDITIONAL RESOURCES

ASTM, 1996, D 517-95 *Guide for the design of groundwater-monitoring systems in karst and fractured rock aquifers*, Volume 4.09 Soil and Rock.

Clark, I., and Fritz, P., 1997, *Environmental isotopes in hydrogeology*, Boca Raton, FL: CRC Press, 328 pp.

Devine, C. E., and McDonnell, J. J., 2005, The future of applied tracers in hydrogeology, *Hydrogeology Journal*.

Fetters, C. W. (2001), Applied hydrogeology, New York: MacMillon Corp.

Flury, M., and Wai, N. N., 2003, Dyes as tracers for vadose zone hydrology, *Reviews of Geophysics*, vol. 41, no. 1, 1002 pp.

Freeman, L. A., and others, 2004, *USGS techniques of water-resource investigations*, TWRI Book 8, Chapter A3.

Kendall, C., and McDonnell, J. J., 1998, *Isotope tracers in catchment hydrology*, Amsterdam: Elsevier, 839 pp.

Nielson, D. M., 1991, *Practical handbook of ground-water monitoring*, Lewis, 717 pp.

Plummer, N., and Friedman, L. C., 1999, *Tracing and dating young ground water*, USGS Fact Sheet 134-99, U.S. Department of the Interior.

6 Aquifer Evaluation

An aquifer test (sometimes called a pumping test) is conducted to evaluate an aquifer by stressing the aquifer, usually through constant pumping, and observing the aquifer's "response" or water-level drawdown in observation wells. Other types of stresses used in aquifer tests include injection or removal of water to cause an instantaneous rise or fall of the water level in a well (*slug* tests). Aquifer testing is a common tool that hydrogeologists use to characterize an aquifer system, confining beds, and flow boundaries.

6.1 HYDRAULIC CONDUCTIVITY

Hydraulic conductivity is a value representing the relative ability of water to move through aquifers of a given permeability material. Hydraulic conductivity can be estimated from lithologic logs (grain size), laboratory measurements, and aquifer tests. Laboratory and field methods are used to determine hydraulic conductivity. However, values obtained in the laboratory are applicable in large-scale situations and may not be representative of the bulk properties of the aquifer. The key value of aquifer tests is that they measure less-disturbed materials.

6.1.1 Grain Size

Hydraulic conductivity (K) for unconsolidated deposits can be estimated from logs of test holes and drill cuttings. Values of K are assigned to the grain size for the material shown in the following list and then multiplied by the thickness of the lithologic unit (Lohman, 1972).

Size: K m/s
Gravel: 3.17×10^{-3}
Coarse sand: 7.05×10^{-4}
Medium sand: 3.52×10^{-4}
Fine sand: 5.29×10^{-7}

6.1.2 Laboratory Measurements

Laboratory determinations from cores of consolidated rocks such as well-cemented sandstone can be used in place to estimate hydraulic conductivity. Reconstituted disturbed samples of unconsolidated cuttings are not representative of field conditions and should not be use for laboratory measurements. It is essential that formation

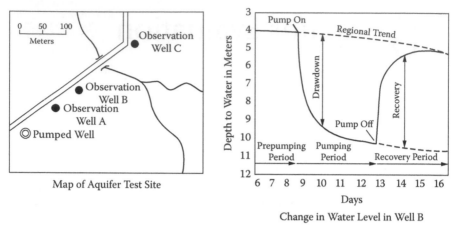

FIGURE 6.1   Diagram of aquifer test. (From Moore, J., *Field Hydrology: A Guide for Site Investigations and Report Preparation*, CRC Press, Boca Raton, 2002. With permission from Taylor & Francis.)

water be used for the laboratory tests. These tests represent only a small volume of the aquifer. An aquifer test, on the other hand, integrates a much larger volume.

6.2   DESIGN OF AQUIFER TESTS

The objectives of an aquifer test are to evaluate the performance of the well, primarily its yield, to estimate its yield and drawdown and to estimate aquifer properties. Accurate estimation of the hydraulic characteristics of aquifers is dependent on reliable data from an aquifer test. Aquifer tests are made in existing wells or in wells drilled for that purpose (Figure 6.1). An aquifer test is a controlled, in situ experiment made to determine hydraulic properties of aquifers (hydraulic conductivity and storage). The test is made by measuring the volume (and rate) of groundwater flow that is produced by pumping a well and observing water-level changes in the pumping wells and/or observation wells. The American Society for Testing and Materials (ASTM) standards and guides for aquifer tests are as follows:

- Standard guide for the selection of aquifer test methods (ASTM, 1991)
- Standard practice for design and installation of monitoring wells in aquifers (ASTM, 1990)
- Standard guide for sampling groundwater monitoring wells (ASTM, 1985)

The following is a list of hydrologic and geologic conditions needed for a successful aquifer test:

- Hydrogeological conditions should not change over short distances or during the duration of the test (no rain).
- No discharging well or stream should be nearby.
- Discharge water should not return to aquifer.

- Pumped well should be completed to the bottom of the aquifer and should be screened or perforated through the entire thickness of the aquifer.
- Observation wells (at least three) should be screened at middle point in the aquifer. One observation should be located outside the area of influence of the pumping well drawdown.
- Location of observation wells should be based on aquifer character.
- Prepumping water-level trend should be determined.

The following are the conditions and field measurements that are needed to take full advantage of an aquifer test:

- Measure water levels before the test, for hours to days, and then just prior to the test and during pumping and recovery.
- Pumped well should be developed adequately prior to test (several hours of pumping and surging).
- A dependable power source should provide a constant pumping rate.
- A flow meter should be capable of reading instantaneous and cumulative discharge.
- Measure electrical conductance, Eh, pH, DO, and temperature.
- Measure water levels measured prior to (days and hours) beginning test.
- Pumping rate should be maintained at 5 percent tolerance. An optimal pumping rate is 50 percent of maximum yield.
- Water-level measurements are made with an electric sounder or pressure transducer.
- It is essential that the discharge water is removed from the site.
- Observation wells should be hydraulically connected to the aquifer being analyzed and should be tested by injecting a known volume of water and measuring the recovery response.
- Baseline trends of regional water-level changes, barometric pressure changes (particularly important for confined aquifers), and local irrigation practices should be established.
- Lithology and construction data for the pumped well should be documented.

6.3   TYPES OF TESTS

6.3.1   SPECIFIC-CAPACITY TEST

The amount of water a well will yield with a certain drawdown can be determined by a specific-capacity test, in which the pumping rate and water level changes are monitored for a set period of time (Figure 6.1). The first step is to measure the initial water level in the well. Commonly, a newly constructed well is pumped at several successively increasing rates for uniform periods (typically 1 hour) to establish a rate that can be maintained for a long-term pumping. The well is then pumped at a steady rate and the water level changes are monitored at the pumped well. Water levels should also be monitored in at least one observation well 2 to 20 meters (6 to 65 feet) from the pumped well. The water level will decline quickly at first as water

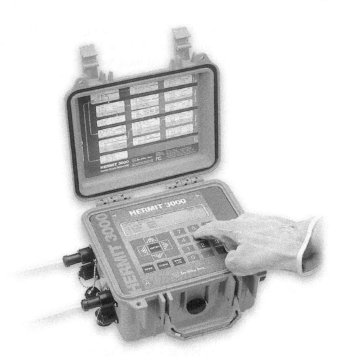

FIGURE 6.2   Instrument Hermit. (Courtesy of In-Situ, Inc.)

is removed from the well, and then more slowly as the rate of flow into the well approaches the pumping rate. The well should be pumped for a sufficient time for the drawdown to either remain constant or approach a constant value. If the test is ended too soon, the calculated specific capacity will be in error. The ratio of the discharge rate (Q) to water level change (drawdown, did) gives the well's specific capacity (Sc), or Sc = Q/did (Figure 6.2). For example, if the discharge rate is 6 liters per second (L/s), or 100 gallons per minute (gpm), and drawdown is 3 meters (10 feet), the specific capacity of the well is 2 L/s per meter (10 gpm/ft) of drawdown. Once specific capacity and the available amount of drawdown are known, the yield of the well can be determined from the formula Q = Sc × did.

The pump should be deep enough that the water level did not go below the pump intake. The pump depth should also be sufficient to allow for drawdown caused by pumping and for natural declines in water level during periods of drought.

6.3.2   Step-Drawdown Test

The step-drawdown test evaluates the performance of the well. Well performance can be affected by resistance to flow in the aquifer itself, partial penetration of the well screen, incomplete removal of mud from the gravel envelope or invasion of fines into the envelope, and blockage of part of the screen area. The well should be developed prior to the test using a surge block and/or pumping until the well discharge is clear. In this test the well is pumped at several (three or more) successively higher pumping rates, and the drawdown for each rate is recorded. The test is usually conducted during 1 day. The discharge is kept constant through each step. The test measures the change in specific capacity. The data provide a basis to choose the pump size and rate for a full aquifer test and for long-term production. The continuous measurements of pH, Eh, DO, temperature, and conductance could provide additional information.

6.3.3 SLUG TEST

In this test a small volume of water is either removed from a well or added to the well bore, and measurements are made of the recovery of the water level. In some cases it is not desirable to add or remove water from the aquifer, for example, at a hazardous-waste site. In these cases a "slug" can be inserted into the well to displace a known volume of water in the well. The slug is quickly placed below the water surface in the well and the subsequent changes in water level are measured and recorded. From the time-drawdown or recovery data, the aquifer transmissivity can be determined. The disadvantages of the test are that a data logger and pressure transducer are needed to measure rapid changes in water levels, storage properties of the aquifer cannot be evaluated, and the test represents only a small volume of the aquifer. Bouwer and Rice's (1976) method of conducting and analyzing data for the slug tests applies to unconfined aquifers, while the Cooper, Bredehoeft, and Papadopulos (1967) method is for confined conditions. The advantages of these tests, as compared with those of full aquifer tests with observation wells, are reduction in cost and time. Many factors contribute to error in slug tests: entrapped air, partial penetration of the aquifer, leaky joints, and the small radius of influence of the test (borehole storage, rather than aquifer properties, may dominate the analysis).

6.4   ANALYSIS OF AQUIFER TEST DATA

6.4.1   THEIS EQUATION

The Theis equation (Theis, 1935) is used to determine hydraulic characteristics of the aquifer. In this analysis a well is pumped and the rate of decline of water level in nearby observation wells (two or more) is noted. The time-drawdown curve is then interpreted to yield the aquifer parameters (transmissivity and storage coefficient). In 1935, Theis developed the first equation to include time of pumping as a factor (Heath, 1963). The assumptions for use of the Theis analysis are as follows:

1. The pumping well is screened only in the aquifer being tested.
2. The transmissibility of the aquifer is constant during the test to the limits of the cone of depression, the depression in the water level that forms around a pumped well.
3. The discharging well penetrates the entire thickness of the aquifer, and its diameter is small in comparison with the pumping rate.
4. The aquifer is homogeneous and isotropic.
5. The aquifer has an infinite areal extent (it's boundaries are beyond the effects of the pumped well).

6. The water removed from storage is discharged instantaneously with decline in head (Lohman, 1972).
7. The pumping rate is constant during the test.

These assumptions are most nearly met by confined aquifers at sites far from the aquifer boundaries. However, if certain precautions are observed, this equation can also be used to analyze tests of unconfined aquifers. Use of the Theis equation for unconfined aquifers involves two considerations:

1. If the aquifer is fine grained, water is released slowly over a period of hours or days, not instantaneously with the decline in head. Therefore, the value of specific yield determined from a short-period test may be too small.
2. If the pumping rate is large and the observation well is near the pumping well, dewatering of the aquifer may be significant and the assumption that the transmissivity of the aquifer is constant is not satisfied.

6.4.2   COOPER–JACOB STRAIGHT-LINE METHOD

The Theis equation is only one of several methods that have been developed for the analysis of aquifer test data. Cooper and Jacob (1946) developed a simplified method that is more convenient to use than the Theis equation. The greater convenience derives partly from its use of semilogarithmic graph paper instead of the logarithmic paper used in the Theis method and the fact that under ideal conditions the drawdown data plot along a straight line rather than along a curve.

6.5   COMPUTER PROGRAMS TO DESIGN AN AQUIFER TEST

6.5.1   CSUPAWE

A computer model program written by Dr. Daniel K. Sunada of Colorado State University in Fort Collins, Colorado, has a designation of CSUPAWE. The program can be used to calculate the water-level response to withdrawal of groundwater. The solution is for a homogeneous, isotropic unconfined aquifer with constant recharge and an initially horizontal water table (Molden, Sundda, and Warner, 1984). The model can be used to determine the best placement of observation wells for an aquifer test and to estimate how long to run the test. The model can also calculate discharge or recharge to the stream in a stream-aquifer system. The model can be run with or without a stream in the system. The program is menu driven and is self-explanatory (Figures 6.3 and 6.4).

Step 1.   Place the supplied program disk in drive A.
Step 2.   Click on CSUPAWE.
Step 3.   The first menu gives the option for selecting flow to a well or flow to a recharge area.
Step 4.   Select flow to a well.
Step 5.   Select Stream in Vicinity or No Stream.

```
 1. Recharge Rate (ft/day).......................................................................2
 2. Transmissivity (ft²/day) ............................................................2500
 3. Specific Yield .............................................................................2
 4. Beginning Time (days) .................................................................30
 5. End of Recharge Period...............................................................0
 6. Beginning Distance (ft) ...............................................................0
    Final Distance (ft)......................................................................500
    Distance Increment (ft)...............................................................50
 7. Depth to Water (ft)......................................................................30
 8. Basin Width (ft) ........................................................................200
 9. Basin Length (ft)........................................................................200
10. Angle from Length Axis (deg) ......................................................0
11. Distance to Stream ...................................................................250
12. Calculate Mound Profile .........................................................yes
13. Calculate Discharge to Stream.................................................yes

Type the number of the variable you wish to change. Type 0 if you wish to continue
without changing.
```

FIGURE 6.3 Computer screen, CSUPAWE. (From Moore, J., *Field Hydrology: A Guide for Site Investigations and Report Preparation*, CRC Press, Boca Raton, 2002. With permission from Taylor & Francis.)

Step 6. Data Input:

Recharge Rate (ft/day): A constant discharge is used.
Aquifer Parameters:
Transmissivity (square ft/day)
Specific yield (dimension less)
Time Period: Calculations are performed at discrete times.

Beginning Time (days) must be greater than 0 and is the first time that the calculations are made. The Time Increment (days) gives the time interval between calculations.

```
 1. Data display
 2. Results display
 3. Graphics display
 4. Results printout
 5. Create file
 6. Another run
 7. Exit
```

FIGURE 6.4 Computer screen. (From Moore, J., *Field Hydrology: A Guide for Site Investigations and Report Preparation*, CRC Press, Boca Raton, 2002. With permission from Taylor & Francis.)

The Final Time gives the last time that calculations are to be made. For example, if the Beginning Time is 10 days, the time increment is 5 days, and the final time is 20 days, then the calculations are to be made at 10, 15, and 20 days. The final and initial times must be integer multiples of the time increment.

End of Recharge Period: The time (days) when artificial recharge or well discharge is terminated. The program will continue to calculate mound profiles until the specified final times.

Distance: Specify the points at which the recharge mound height or water-level decline is to be calculated. The Beginning Distance (ft) is always set at 0, which is directly under the center of the basin or at the well. The distance increments (ft) give the distance between points of mound height calculations. The Final Distance (ft), the last point at which the mound height is to be calculated.

Depth to Water (ft): The depth to water is the distance from the ground surface to the water table in feet.

Saturated Thickness (ft): This is the distance from the bottom of the aquifer to the initial water table.

Basin Geometry: A rectangular basin is used, so enter the Basin Width and Basin Length in feet.

Angle (degrees): This specifies the angle that a vertical plane makes with a line drawn from the center of the well or recharge basin.

Distance to Stream or Distance to Impermeable Boundary (ft): The distance to the boundary is measured from the center of the recharge basin or well to the stream.

Calculate Mound Profile and Calculate Discharge to Stream: You can have the program calculate either the mound profile or the discharge to the stream or both.

Change in Variables: Variables are changed by typing the number corresponding to the variable. For example, if you want to change the F, enter 2. The old value of the variable will appear on the screen and you will be asked to input the updated value. After the updating is completed, you will be returned to the main data display. When you press just the Enter key, the program begins calculating with the current parameters.

Step 7. Calculations: As points are calculated they are plotted on the screen. When the calculations are finished you will be asked to press Enter to continue. All results are kept in memory, and the Output Options menu appears.

6.5.2 THEIS COMPUTER MODEL PROGRAM

Another computer program prepared by John McCain of the USGS is called THEIS. It is simpler than CSUPAWE. The THEIS program solution calculates drawdown due to a pumping well at a constant rate. The input and output can be in metric or English (U.S.) units. The program is distributed as freeware.

6.6   CONSTRUCTION OF HYDROGEOLOGIC
MAPS AND CROSS SECTIONS

The major maps and cross sections constructed by hydrogeologists are water-table contour, saturated thickness, bedrock contour, water-level change, and transmissivity. The water level contour map shows the elevation of the water table for a specific hydrogeologic unit. Depth to water data is first converted to water-level elevations. Water-level measurements should be measured within a few days of each other because the map represents a specific point in time. Elevations of water levels in streams, lakes, and ponds can be used to construct the map if they are in hydrologic connection with the underlying groundwater. After the water-table/water-elevation points are plotted, the contour lines are drawn and smoothed. Avoid bull's-eye contours that have only a single point for control. The water-table contour maps are valuable tools for understanding the groundwater flow system by providing information on gaining and lowering reaches of the stream, no-flow and flow boundaries, and areas of groundwater withdrawal.

6.7   HYDROGEOLOGIC SECTIONS

Construction of a hydrogeologic section is used to visually depict the hydrogeology. A section is a two-dimensional view of the subsurface. Hydrogeologists use sections to understand water flow underground and make estimates about boundaries to groundwater flow.

ADDITIONAL RESOURCES

ASTM, 1985, D 4448 *Standard guide for sampling ground water monitoring wells.*

ASTM, 1990, D 5092 *Design and installation of ground-water monitoring wells in aquifers.*

ASTM, 1991, D 4043 *Standard guide for selection of aquifer test method in determining of hydraulic properties by well techniques.*

ASTM, 1993, D 4750 *Standard test method for determining subsurface liquid levels in a borehole or monitoring well (observation well).*

ASTM, 1995, D 5717 *Guide for the design of ground-water monitoring systems in karst and fractured rock aquifers.*

ASTM, 1996a, D 5730 *Standard guide for site characteristics for environmental purposes with emphasis on soil, rock, the vadose zone and ground water.*

ASTM, 1996b, D 6000 *Standard guide for presentation of water-level information from ground-water sites.*

Bentall, R., 1963, *Methods of collecting and interpreting ground-water data, examination of drill cuttings,* U.S. Geological Survey Water-Supply Paper 1544-H.

Bouwer, H., and Rice, R. C., 1976, A slug test for determining hydraulic conductivity of unconfined aquifers with completely or partially penetrating wells, *Water Resources Research,* vol. 12, 423–428 pp.

Carrillo-Rivera, J. J., Cardona, A., and Moss, D., 1996, Importance of the vertical component of groundwater flow a hydro geochemical approach in the valley of San Luis Potosi, Mexico, *Journal of Hydrology,* vol. 185, no. 1–4, 23–44 pp.

Clark, I. D., and Fritz, P., 1997, *Environmental isotopes in hydrogeology*. Boca Raton, Florida: CRC Press, 328 pp.

Cooper, H. H., Bredehoeft, J., and Papadopulos, I. S., 1966, Response of a finite-diameter well to an instantaneous charge of water, *Water Resources Research*, vol. 3, 263–269 pp.

Drescher, W. J., 1961, Aspects of ground-water investigations as related to contamination, in R. A. Taft, (Ed.), *Sanitary engineer*.

Garber, M. S., and Koopman, F. C., 1968, *Methods for measuring water levels in deep wells*, in *USGS techniques of water-resource investigations*, TWRI Book 8, Chapter A1, 32 pp.

Heath, R., 1963, *Basic ground-water hydrology*, U.S. Geological Survey Water-Supply Paper 2220, 84 pp.

Kendall, C., and McDonnell, J. J., 1998, *Isotope tracers in catchment hydrology*. Amsterdam: Elsevier, 839 pp.

Lohman, S. W., 1972, *Ground-water hydraulics*, U.S. Geological Survey Professional Paper 708, 70 pp.

Mercer, J. W., 1991, *Methods for subsurface characterization*, U.S. Environmental Protection Agency.

Molden, D., Sunada, D. K., and Warner, J. W., 1984, Microcomputer model of artificial recharge using Glover's solution, *Ground Water*, vol. 22, no. 1, 73–79 pp.

Moore, J. E., 1991, *A guide for preparing hydrologic and geologic projects and reports*. Kendall/Hunt Publishing Co., 96 pp.

Nielsen, D. M., 1991, *Practical handbook of ground-water monitoring*. Lewis Publishers, 717 pp.

Taylor, C. J., and Alley, W. H., 2001, *Ground-water-level monitoring and the importance of long-term water-level data*, U.S. Geological Survey Circular 1217, 68 pp.

Theis, C. V. (1935). The relation between the lowering of the piezometric surface and the rate and duration of discharge of a well using ground-water storage, *American Geophysical Union Transactions*, 16, 519–524 pp.

7 Streamflow Measurements

Stream gauging is the measurement of discharge in an open-water channel. Gauging of streamflow has a practical application to the design of dams, flood control structures, surface-water allocation, and relation of surface water to groundwater. Stream rating curves are constructed by plotting values of gauged streamflow against the stage of the stream for a given flow. The rating curve is used to predict baseflow recession.

7.1 BASIC EQUIPMENT

The basic equipment for performing discharge measurements is a current meter, a timer, sounding equipment, and a tagline. A current meter consists of an impeller, which, when placed in the stream, rotates in proportion to the speed of water passing the instrument. The rate at which the impeller rotates is recorded by an electronic counter that is heard on earphones or a counter or acoustic doper.

7.2 MEASUREMENT PROCEDURE

The stream discharge is measured as follows:

1. Select a straight reach of stream with a smooth shoreline.
2. Extend a tape (tagline) across the stream.
3. Develop a water depth profile.
4. Divide the stream into segments at regular intervals.
5. Take the measurements 10–20 stations along a profile.
6. Clamp the meter at 0.2 and 0.8 of total depth.
7. Take a velocity reading at each station.
8. Compute discharge: cross-section area × velocity = discharge for each section. Compute the total discharge by adding the discharges for each section.

The area of flow is computed from observations of width and depth. For wading measurements, the width is generally marked by a tagline. The depth of the stream is measured at a sufficient number of intervals across the stream section to adequately determine the total area. For wading measurements, depths should be read to hundredths of a millimeter where practical.

Difficulties in obtaining accurate velocity observations will often arise that will tax the ingenuity of the hydrologist. These difficulties arise from field conditions

such as soft streambeds, scour under the wading rod, sand dunes, boulders, and swift
or extremely slow-flowing or very shallow water.

Shallow depths and low velocities in streams render current-meter measurements
inaccurate, if not impossible to obtain. Under these conditions a weir or flume may
be useful for obtaining the discharge measurements.

Stream gauging poses hazards that threaten the safety of the hydrologist. Personal
safety depends upon a healthy respect for rivers and streams; in the words of
Shakespeare's Falstaff: "Discretion is the better part of valor."

7.3   METHODS TO DETERMINE STREAM–AQUIFER RELATIONS

Several field techniques can be used to determine the relation of the aquifer and a
stream. The most commonly used ones are as follows:

- Stream discharge (flow) measurements at a series of points along a stream
  during a relatively short period and during relatively constant flow condi-
  tions (called *seepage runs*)
- Stream gauging
- Tracer studies
- Water-table contour maps

Stream channel gain and loss studies can be made to evaluate the effect of groundwater withdrawal on the stream flow. Surface-water measurements were used to
evaluate the effect municipal wells withdrawal on the flow of the Arkansas River
at Lamar, Colorado (Moore and Jenkins, 1966). The study showed that municipal
pumping had lowered the water table below the streambed, the hydraulic connection
between the stream and the water table was broken, and an unsaturated zone had
developed below the streambed near the area of greatest pumping. The channel loss
data were used to compute the vertical hydraulic conductivity of the streambed.

Streamflow hydrographs can be analyzed to estimate the groundwater component of streamflow. The groundwater component of streamflow is termed *base flow*
(Winter and others, 1998).

Environmental tracers can be used to determine the source areas of water, water
age, and chemical reactions that take place during transport (Winter et al., 1998).
Useful tracers include dissolved constituent (cations and anions), isotopes of oxygen, tritium, radon, and water temperature. Several sections along the river will be
required.

Water-table contour maps can be analyzed to determine if the stream is gaining
or losing water from the aquifer. Contours of water table elevation indicate gaining
streams by pointing in an upstream direction, and they indicate losing streams by
pointing in a downstream direction in the immediate vicinity of the stream (Hurr
and Moore, 1972).

ADDITIONAL RESOURCES

Buchanan, T. J., and Somers, W. P, 1969, *Discharge measurements at gaging stations*, in *USGS techniques of water-resources investigations*, TWRI Book 3, Chapter A8, 65 pp.

Hurr, R. T., and Moore, J. E., 1972, *Hydrologic characteristics of the valley-fill aquifer in the Arkansas River Valley, Bent County, Colorado*, U.S. Geological Survey Hydrologic Atlas, 461 pp.

Moore, J. E., and Jenkins, C. T., 1966, An evaluation of the effect of groundwater pumpage on the infiltration rate of a semipervious streambed, *Water Resources Research*, vol. 2, no. 4, 691–696 pp.

Winter, T. C., J. W. Harvey, O. L. Franke, and W. M. Alley, 1998, *Ground water and surface water a single resource*, U.S. Geological Survey Circular 1139, 79 pp.

8 Hydrogeologic Reports

The report is the most important aspect of many geologic and hydrologic projects, because it is normally the only product of the project that the client will see. A properly planned, well-written, and carefully reviewed report enhances the credibility of the project investigator and the organization (Figure 8.1). The report is written for other specialists and/or educated layman. The technical report may have to be reorganized to make it easier to use for some readers.

This chapter describes critical steps in the planning, writing, and review of hydrologic, geologic, and related reports. The chapter includes guidelines, checklists, and source references to assist authors in the various stages of report preparation. The information is a suggestion and is not intended as a replacement for report formats established by your organization or supervisor. The objective of these steps is to help produce quality reports on time (Figure 8.2).

This guide consists of three basic segments.

Report Planning—discusses the use of report work plans and outlines
Report Writing—describes writing the report title, executive summary or abstract, introduction, body, and summary
Report Review—describes procedures to be followed during editorial and technical review of a report

8.1 REPORT PLANNING

Report planning should precede report preparation (Figure 8.3). Some suggestions for report planning are

- Prepare a topical outline during the first week of the project.
- Prepare an annotated report outline during the first one third of the project (e.g., first month of a 3-month project).
- Write interim reports to reduce the size and complexity of the final report. Interim reports can be useful for describing the preliminary results of analyses and preliminary interpretations during the duration of the project.

Report work plan. The report work plan should include projected dates for completion of outlines, report writing, first draft, review, approval, and release. One third of the life of the project is typically devoted to report writing, review, and approval.

- Think about the report topic and list ideas that come to mind (brainstorm)
- Write a rough draft on a word processor
- Print a copy of the draft and mark it up
- Revise the draft
- Print a copy of the draft and mark it up
- Have a colleague review the draft
- Write a final draft
- Proofread and correct

FIGURE 8.1   Report writing process. (From Moore, J., *Field Hydrology: A Guide for Site Investigations and Report Preparation*, CRC Press, Boca Raton, 2002. With permission from Taylor & Francis.)

Report writing guides. Writing is a continuing effort throughout the duration of the project and should never be treated as a chore to be left until the end of the project. The writing of a report will be successful if careful planning precedes the effort. The following are guides for report writing:

- Define the audience of the report.
- Find published reports that could serve as models.
- Prepare a topical outline and have it reviewed.
- Prepare an annotated outline and have it reviewed.
- Send a copy of the outline to the client for review.
- Write the report background, purpose and scope, description of study area, methods of study, and review of previous studies.
- Write each paragraph of the first draft and put each paragraph on a separate page.

- Logical organization where more important items stand out
- Writing style fits the intended audience
- Minimal jargon
- Effective illustrations
- Clear simple tables
- Pleasing design (cover page)
- Pleasing and appropriate layout

FIGURE 8.2   Attributes of a quality report. (From Moore, J., *Field Hydrology: A Guide for Site Investigations and Report Preparation*, CRC Press, Boca Raton, 2002. With permission from Taylor & Francis.)

A through and competent review is essential to maintain the technical quality of reports. The purpose of the review is to give a technical evaluation that will improve the report and eliminate errors that would embarrass the author and his agency. The following are guidelines and procedures for technical reviews:

- *Technical correctness* – Is the report technically valid? Are conclusions properly supported by correctly interpreted data? Are assumptions reasonable and clearly stated?
- *Readability* – Is it written for the intended audience, with correct grammar, syntax and a minimum of scientific jargon? Are illustrations and tables legible and understandable?
- *Title* – Is it explicit and does it reflect the objectives of the report?
- *Abstract* – Does it state the purpose of the report? Does it describe the study and summarize the results and conclusions?
- *Introduction* – Does it clearly describe the problems addressed in the report, state objectives, and scope of the report?
- *Methods* – Were appropriate techniques used in the study?
- *Conclusions or results* – Do they summarize the principal finding of the study and answer each of the objectives described in the introduction? Are they sound and properly documented?
- *References* – Are all references cited in text included in this section? Are they complete and cited correctly?

FIGURE 8.3 Instructions for technical reviews. (From Moore, J., *Field Hydrology: A Guide for Site Investigations and Report Preparation*, CRC Press, Boca Raton, 2002. With permission from Taylor & Francis.)

- Revise the first draft. It usually takes several drafts (four or five) to get a report to the stage at which it is ready for initial review by someone other than the author.
- Have the report edited before peer (technical) review.
- Arrange for concurrent peer reviewers.
- Respond, in writing to peer-review comments.
- Have the report edited again, if many revisions are made in the peer review.
- Complete work on and print the report, and deliver to the client as quickly as possible.

Generic report outline. The following is a generic outline of a technical report:

1. Title
2. Executive Summary or Abstract
3. Introduction. Statement of problem or information needed
4. Purpose and Scope
5. Body of Report. Study approach, methods of data collection and analyses, results of those analyses, or answer to the problems stated in the introduction
6. Summary and/or Conclusions
7. References

Report organization. The organization of a report requires the author to make decisions on the content and order of presentation of topics. Each report presents a different problem in organization, and no cookbook method of report organization can be given. The report title, contents, purpose and scope, and summary and/or conclusions must be consistent. The first step in report organization is the preparation of an outline. An outline helps authors to organize their thoughts and plans early in the project and to focus project activities throughout the life of the project.

After selection of the title (see following section), authors should prepare a topical outline that contains major and minor headings that reflect the title and organization of the report. It is far easier to reorganize an outline than a completed report.

The next step is to prepare an annotated outline. This outline should be as detailed as possible. The annotated outline is an expanded version of the topical outline. A topic (lead-off) sentence or paragraph is prepared for each heading in the topical outline. After the annotated outline is prepared, it needs to receive the same review as the topical outline. The annotated outline may be sent to the client for review. This early review helps ensure that the final report will meet the needs of the client.

8.2   REPORT WRITING

8.2.1   Overcoming Writer's Block

Putting down the first word is hard for some writers, including many who have written best sellers. Writer's block and procrastination are closely related. The authors who have writer's block must put down that first word or sentence, good or bad, and use self-imposed control to finish the report. Some authors begin by writing the description of the study area or previous investigations. One way to develop an outline for the report is to randomly write down your main thoughts (bullet format) of what you want to say. Then group your thoughts that are related and arrange them in outline form.

8.2.2   Title

The title of the report should be brief but informative. It is a concise description and perhaps even the ultimate abstract of the subject of the report. The title of the report will be read by many more readers than the report itself. One of the best methods to ensure that the report will reach the intended audience is through a complete title that can be properly indexed for bibliographic files. Ideally, the title should be as short as possible, yet contain the following:

- Subject(s) of report (states the problem or issue addressed in the report)
- Location of study area (if appropriate)
- Time or period of study (if appropriate)

For example, the report title "Potentiometric Surface of the Floridian Aquifer in Southwestern Florida, October 1988" includes all the above elements.

The title should not contain abbreviations or jargon. Acronyms are permitted only if the source words are first spelled out in the title and the acronym is needed to describe the subject of the report. Word names for computer programs also are permitted if the title clearly defines their meaning.

Some examples of weak titles and their suggested revisions are as follows:

Weak: HYDRAUX: A One-Dimensional, Unsteady, Open-Channel Flow Model
Revised: The Computer Program HYDRAUX, a Model for Simulating One-Dimensional, Unsteady, Open-Channel Flow
Weak: Integration of Computer Associates Telegraph and Text Editors
Revised: Integration of Computer Graphics and Text-Editing Software for Production of Reports
Weak: Colorado Uranium
Revised: A History of the Development of Uranium Mining in Colorado

8.2.3 EXECUTIVE SUMMARY OR ABSTRACT

A well-prepared executive summary (expanded abstract) or abstract captures the basic content of the report. This section should contain the following parts:

- An opening statement that includes the reason for the study, its scope, and a statement of cooperation, if any. For example: "This report describes the results of a study by Dames and Miller done in cooperation with _____ to evaluate _____."
- Type of study if not evident from report title; for example, water-quality study, case history, hydrologic reconnaissance, progress report, original research, aerial investigation.
- Results of study, in decreasing order of importance.
- Conclusions, if any.

The executive summary or abstract is prepared after the paper is written and should not contain information that is not included in the report. Although there is no general word limit rule, a 250-word limit is imposed by many journals for abstracts. An executive summary is commonly 750–1,000 words long.

The inclusion of reference citations, abbreviations, and acronyms in summaries should be avoided. Note that some journals do not require that abbreviations be spelled out where first used. Authors should be aware of publishers' guidelines when preparing executive summaries or abstracts for journals and symposia.

The following are examples of weak executive summaries or abstracts and their revisions:

Weak: This report describes a computer-model program that simulates conditions in the water-table aquifer of the Pine Barrens in southern New Jersey. The model simulates seepage from the aquifer to streams and swamps. Groundwater flow is approximated in two dimensions. Theoretical development of equations is presented as well as documentation of input data and instructions for use of the model.

Revised: A preliminary, two-dimensional, steady-state model of the water-table aquifer underlying the Mullica River basin, Pine Barrens in New Jersey, was constructed as an initial step in developing a predictive model. The purpose of the model is to evaluate concepts of the flow system and data required to simulate it. The computer model is based on the finite-difference method for solving stream-seepage equations coupled to the groundwater-flow equation. Model-simulated water levels and streamflow compare closely with measured values. Initial estimates of streambed hydraulic conductance and aquifer hydraulic conductivity were adjusted to those used in the model. Simulated water levels were within 5 feet of measured water levels at 41 of 42 wells. Simulated streamflow was within 20 percent of measured streamflow at 12 of 15 sites.

Weak: This report contains information about the occurrence, quality, quantity, and direction of movement of groundwater in Hampton County.

Revised: During an investigation of groundwater hydrology in Hampton County, Nebraska, water levels were measured in 196 wells and water samples for chemical analysis were collected from 188 wells and springs, mainly during September 16–27, 1974. Fifteen wells and 3 springs were resampled in March and May 1975. The dissolved solids in the samples ranged from 150 to 300 mg/l. The groundwater movement is toward the stream, which is the discharge point for the sand and gravel aquifer.

8.2.4   INTRODUCTION

The introduction begins with a brief discussion of the need for the study, that is: Why was the study done? The introduction also includes brief description of the study area and a statement of cooperation, if applicable. A statement of the purpose and scope of the report follows this introduction.

8.2.5   PURPOSE AND SCOPE

The purpose of a report needs to be related to that of the overall study. For example, the purpose (objective) of a report might be to describe long-term trends in concentration of chloride within a study area, whereas the purpose of the study might be to describe the overall hydrogeology of the study area. The scope of a report might include, for example, the time period analyzed, the number of samples collected and analyzed, the database used, and the analytical techniques.

- Description of study area—This section contains a brief discussion of the location and size of the study area; its climate; and physiographic, geologic, and hydrologic (or hydrogeologic) setting. Other descriptive information on the study should be given only if it is pertinent to the objective(s) of the report.

- Methods of study—The methods section contains a description of methods used. Remember, the heading "Methods of Study" absolutely limits the content of the material that follows.
- Approach—The "Approach" section differs from the methods section by presenting the rationale behind the study and the manner in which the study was performed. For example, a statement of an approach might be as follows:

  The study involved three phases of activity: (1) Organizing and evaluating the geohydrologic data in order to develop a conceptual model of the groundwater basin of the San Bernardino Valley, (2) developing a steady-state and transient-state digital-computer model of the basin, and (3) using the computer model to predict groundwater levels under selected pumping alternatives, primarily in the artesian areas of the basin.

- Previous studies—The previous studies section acknowledges past work encountered as part of the project literature review. Be sure that this information is accurate and as complete as possible. Be sure to compile all bibliographic information for all citations (here and in the rest of the report) at the time the source publication is examined to avoid time-consuming bibliographic searches after the report is written.

8.2.6 BODY OF REPORT

The body of the report contains the data and interpretations that answer the problems stated in the introduction. The body of the report presents information in the form of text, illustrations, and tables. The author needs to develop all discussions along the main theme of the report as noted in the title, contents, introduction, and purpose and scope.

8.2.7 SUMMARY, CONCLUSIONS, AND RECOMMENDATIONS

These sections provide a solution to the problem(s) stated in the introduction, focus on the significant findings, and recommend a further course of action.

The most widely read parts of a report are the executive summary or abstract and the summary, conclusions, and recommendations, because these sections state in a condensed form the most important ideas and findings, and their significance. The executive summary or abstract and concluding material must be in complete agreement and must present the essential information in the report. They should not be mere repetitions of one another, although the same statements and data can be included in both. Guidelines for preparing these sections are given below.

A summary is a restatement of all the main ideas presented in the report and generally is required for all interpretive reports. The purpose of a summary is to recapitulate the most important facts so that the reader can correctly recall the results and their significance. The summary should describe or list these items in the order in which they are presented in the text; to do this, the author needs to review the table of contents and the main discussions when writing the summary.

The conclusions and/or recommendations sections state the final results, interpretations of the study, and recommendations for further action. All conclusions must be stated either in the report or be able to be readily derived from the material presented therein. In preparing the conclusions, the author should refer to the "Purpose and Scope" section to verify that the sections support each other—that is, the stated purpose of the study has been fulfilled, the scope has been adhered to, and the purpose and scope are reflected in the body of the report and in the conclusions. In general, the conclusions are listed in the same order as the corresponding objectives in the "Purpose and Scope" section. The main conclusions also should be incorporated in the executive summary or abstract. The recommendations state what further actions, if any, should be taken to address or resolve the problem(s) stated in the introduction, and the potential consequences or benefits of such actions.

8.3   REPORT REVIEW

The objective of a competent and thorough review (Figure 8.3) is to

- ensure the report achieves the goals stated in the proposal,
- ensure the readability of the report,
- ensure the technical quality of the report,
- evaluate suitability of the proposed publication media,
- evaluate the effectiveness of the presentation, and
- correct errors and deficiencies that could embarrass the author and/or the parent organization.

The peer/technical reviewer should evaluate the organization of the report. Are the major headings in the table of contents reflected in the title and purpose and scope, and do the summary or conclusions address the objectives stated in the purpose and scope? The peer reviewer should determine if the report is free of technical errors

The reviewer should be objective, direct, careful, reasonable, and considerate. The following guidelines are provided to reviewers to improve the quality of their review:

- *Be objective.* Examine your attitude carefully before you begin a review, and examine your attitude at frequent intervals during the review. Are you trying to show how smart you are? There is no place for sarcasm in the review comments. Remember, being a reviewer is a position of trust.
- *Be direct.* You should avoid vagueness and ask clear questions. If there isn't room to make intelligent questions or comments, use a separate sheet of paper. Isolated question marks are not acceptable forms of inquiry.
- *Be careful.* The author and supervisor have a responsibility to ensure that the report is as free from errors as they can possibly make it. Most editorial and

technical errors should be eliminated before the report is submitted for technical review. Very little time should be spent by the reviewer in editing.

- *Be reasonable.* Constructive suggestions are appreciated. Remember, however, that when the report is sent for colleague review, most of the allotted time and money has already been spent.
- *Be considerate.* Place yourself in the shoes of the author. Refrain from the use of humor, witticism, and sarcasm in your comments. No matter how funny it seems to you at the moment, you can be sure the author will misunderstand and be resentful of even the most well-meaning barbs. Be brief and courteous in your remarks. Remember that your report will be reviewed, and you should treat your peer's report as you would like yours treated.

8.3.1 PEER REVIEW

A peer reviewer is, next to the author, the most important member of the report team. The ability to do a good technical review is learned by practice. The objective of the peer review of the report is to ensure its technical soundness and to help the author improve the report. The following is a list of responsibilities for peer reviewers:

- Identify technical aspects of a report for they do not have the necessary expertise to review and convey this information to the author so that additional reviewers can be selected.
- Show willingness to review the report in a timely manner.
- Ensure technical soundness and clarity, and suggest alternative methods of analysis or interpretation, if appropriate.
- Devote adequate time and effort to check the mathematics, approach, organization, editing, adequacy of data used to support conclusions, applicability and soundness of illustrations and tables, and readability.
- Clearly indicate problems, describe any in the report, and prepare a summary of the review.
- Make all constructive suggestions for improvements. Reviewers should point out both positive and negative aspects of the report.
- Communicate with the author during the review process.
- It will be helpful if all calculations are on a separate sheet.

8.3.1.1 Review Methods

There are four basic methods of peer review: concurrent, consecutive, group, and storyboard. The method most commonly used is concurrent review because it consumes the shortest period of time. Group and storyboard reviews are similar and are useful in situations in which the author needs special assistance.

Concurrent review. Copies of the report are sent to each reviewer simultaneously and all comments are incorporated at one time. If reviewers disagree on a particular point, the author sends each a set of the other's review comments and asks them to resolve the conflict and notify the author of the resolution. If the conflicts are minor, they might be resolved by a telephone conference call. In any instance, the author is required to document all correspondence with reviewers, whether written or oral.

Consecutive review. Copies of the report are sent to a reviewer, and after the corrections have been made, the report is subsequently sent to a second reviewer. An advantage of this type of review is that the report, in theory, is improved after each review, assuming that the second reviewer will detect errors missed by the first reviewer (and that the author does not introduce new errors in the process of responding to the previous reviewer).

Group review. Two or more colleague reviewers are sent copies of the manuscript. After they have completed their review, the reviewers and the author meet to discuss and revise the report. Commonly, these meetings bring group interaction and discussion that result in a greatly improved report. A facilitator might be needed to control the pace of the meeting and to mediate any conflicts that arise.

Storyboard review. Mockups of the text and illustrations are displayed in sequence on a table or wall. A blank sheet of paper can be attached to the pages for reviewers' comments. After the review, the author compiles the comments and discusses them with the reviewers. This type of review is especially well suited for map and stop-format reports.

8.3.2 EDITORIAL REVIEW

The editorial review of text, illustrations, tables, and the manuscript package should consider consistency in the use of terminology, clarity of expression, proper grammar, agreement of content with headings and figure titles, adherence to publisher's style, consistent use of topic sentences for paragraphs, completeness of all components and support documents, suitability of illustrations, and readability. A description of key steps in the editorial review of the text, illustrations, tables, and manuscript is given below in the form of questions.

Text
- Is the format of the report appropriate for the intended outlet?
- Is the title of the report concise and accurate, and does it follow policy?
- Are the title and authorship the same on the cover title page, abstract or executive summary page, and transmittal memorandum?
- Has assistance given by outside sources been acknowledged?
- Illustration list—Is the type of illustration indicated? Is the caption correctly abridged? Does it avoid abbreviations and acronyms?
- Tables list—Does the title of each table match the table title or a correctly abridged version thereof? Does it avoid abbreviations and acronyms?
- List of conversion factors and abbreviations—Does it include all units of measure in the text, illustrations, and tables, and no others? Is the format correct?
- Has the entire report been read for grammatical and spelling accuracy and for internal consistency, preferably before peer review and again before submission to management?

- Is the wording clear and free of jargon?
- Do text headings agree with the table of contents in wording, level, and page number? Do discussions under a heading pertain to the heading?
- Units of measure—Is the international system (metric) or Imperial system (inches, pounds) among others, used consistently? (Note that it is acceptable to mix inch-pound units with metric units of chemical concentration and specific conductance.)
- Do numbers and descriptive material in the text agree with the data in tables and with information shown in illustrations?
- Are all bibliographic citations in the text, tables, and illustrations in the list of references? Are they in the correct format, and do authorship and year of publication agree with information in the list of references?
- Are figure and table cut-ins placed in the text or in the margins where they are first mentioned in the text?
- Are figure and table caption pages immediately following the page containing the initial reference?
- Do the summary and/or conclusions contain only information given in the report, and do they answer the purpose(s) of the report?
- Are the recommendations reasonable for information given in the report?

Illustrations

- Is the format of the illustration correct?
  - Does the format meet the standards of the publisher?
  - Is the format of similar illustrations the same?
  - Is the explanation, if needed, complete and in the proper format?
  - Is the illustration caption correct?
  - Does the caption reflect the figure's content?
  - Is the source of the illustration cited at the end of the caption, and is the number of the figure in the cited source included in the citation?
  - Are the illustrations properly cross-referenced in the text?
  - If the illustration is a graph:
    - Are axis labels, grid, scale, and units of measurement appropriate?
    - Are the axes properly labeled?
    - Are necessary geographic names given in the figure captions?
    - Does the title of the graph reflect information depicted by the graph?
  - If the illustration is a map:
    - Does the map contain all geographic names and sites in the study area that were mentioned in the text?
    - Are the coordinate systems properly presented?
    - Is the explanation correct? Are all symbols, physical features, and abbreviations labeled and explained?
    - If the map is a plate, is the plate caption complete, including the type of illustration and complete geographic location?
    - Is a base-map and/or a data-credit note needed?

- – If colors are used, are they the same on the map and sections?
- – Are data contours explained and are datum notes needed?
- – Is information in the map in agreement with data mentioned in the text, presented in a table, or shown in another illustration?
- – Are the scale and a north arrow included?
- If the illustration is a cross-section:
  - – Are the same vertical and horizontal scales used, and if not, is the amount of exaggeration shown?
  - – Is the view from the east or south?
  - – Are the end points labeled and do they correspond to those shown on the map trace?
  - – Are the maps that show the section traces referred to?
  - – Is it possible to use the same horizontal scale as that on the map showing the section traces?
  - – Are the intersections with other sections identified?
- Are all symbols, physical features, and abbreviations explained?

Tables
- Are the data accurate?
- Are the data presented logically?
- Were the data in the table checked against the data mentioned in the text, presented in an illustration, or presented in another table?
- Are the tables properly cross-referenced in the text?
- Is the position of the table in the report appropriate?
- Does the table follow the first principal reference to the table?
- If a table is long, should it be moved to the back of the report, perhaps as an appendix?
- Is the format of the table correct?
- Does the format meet the standards of the publisher?
- Is the format of similar tables the same?
- Are head notes and footnotes used properly?
- Does the presentation of data in the table parallel the table title and the discussion of the table in the text?
- Are the location of the data and period of record needed in the table title to understand the table?
- Is the source of the table or data cited (in a head note)?
- Is the number of significant figures presented correctly and in a consistent manner?
- Is an unnumbered table properly introduced?
- Are all geographic names and sites in a table located on a map?

8.3.3 TECHNICAL REVIEW

The importance of technical review in the preparation of quality reports cannot be overemphasized. It is suggested that at least two reviewers be used for all reports. The reviewers should be selected on the basis of their special knowledge, experience, or interest in the subject material in the report. Where practical, it is desirable that in the case of interpretive reports, at least one technical reviewer be selected from outside the originating office.

Technical review of a report may reveal shortcomings or omissions that cannot be corrected in the final stages of the report. To avoid this occurrence, it is essential that the reviewer(s) be selected at the beginning of the project and be informed and consulted during the project to allow incorporation of methods, actions, and so forth at the appropriate time, as necessary for the completion of a satisfactory report review. Compromise in report quality or substantial delays and extra costs may result if this course of action is not followed.

A technical reviewer should concentrate on the technical adequacy of the report, but any major editorial errors, particularly in organization, should be pointed out. Reviewers should summarize their comments and make recommendations for improvement of the report in a memorandum to the author. Brief, clear, and legible review comments should be entered directly on the manuscript. The author should respond in writing to accept or reject all comments from reviewers.

Reviewers should adopt a systematic approach when evaluating reports. The 14-step method given in Section 8.4 provides a system for performing technical reviews.

8.4   REVIEW STEPS

The subsequent 14 steps, if followed, can greatly improve the quality of peer review. Although not included in the steps, peer reviewers should attempt, as part of their review, to evaluate the editorial quality of text, tables, and illustrations as described in Section 8.3.2.

1. Transmittal Memorandum. Carefully read the transmittal memorandum from the originating office and other background information on the project that generated the report. Such information can be extremely helpful by indicating the emphasis needed to be placed on various parts of the report.
2. Title. Carefully read and study the report title. The title should convey the subject(s) of the report yet be as short as possible. More than 15 words might be too many.

   - Does the title accurately reflect the main theme of the report and first-level headings in the contents?
   - Is the location of the study area included?
   - Is it necessary to include the time period of the study or of the data set?
   - Does the title avoid abbreviations, acronyms, brand names, and extraneous words?

3. Contents. Carefully examine the table of contents—it tells the reader the order in which the topics are discussed and indicates the relative importance of these topics. A well-organized table of contents shows that the author has written a report with a logical and orderly presentation of information.

- Do the first-level headings in the main body of the report accurately relate to the key words in the report title, both in wording and in order of presentation?
- Are the contents logically organized in a manner that contributes to continuity of thought?
- Are all headings appropriately subdivided so that the subheadings further develop the subject of the heading?
- If subheadings are listed under a heading, are at least two subheadings used?

4. Conversion Table. If the report contains a unit-of-measurement conversion table, compare it with units of measurement in the text, illustrations, and tables of the report.

- Does the table include all units of measurement used in the text, illustrations, and tables, and no others?
- Are the units of measurement worded correctly, and are the abbreviations in the form required by the publisher?
- Does the table include a definition of *sea level* if used in the report?

5. Abstract or Executive Summary. Read the abstract or executive summary several times. This section is a digest of the information in the report. It can be written only after completion of the entire report.

- Is this section of an interpretive report informative rather than indicative?
- Does it reflect the summary and/or conclusions and stress the most important results in order of decreasing importance?
- Does it contain information that the reader can readily find in the body of the report?
- Does it address the purpose of the report and include the most salient findings of the report in decreasing order of importance?

6. Introduction. The introduction sets the theme of the report and establishes the logic of the presentation that follows. It also is a place for miscellaneous information that does not belong in the body of the report.

- Does the introduction clearly define the need for and the purpose of the investigation—that is, the what, why, where, and when of the investigation? Does it relate to the main theme of the report as indicated in the title and table of contents?
- Does the purpose and scope of the report define the objectives of the report and reflect the title and table of contents? Does it pertain only to the report (not to the project itself)? Does the scope of the report describe the depth of discussion in developing the subject of the report?

- Are the methods or approaches or both stated briefly, and are they appropriate to the problem and purpose of study? Does the methods section pertain only to methods? Remember that new methods and approaches will need more detailed explanations than will standard methods and approaches.
- Does the introduction describe the physical setting of the project area, giving only that information necessary to understand the data and interpretations?
- Is previous work in the subject area discussed and properly referenced?
- Is information and assistance obtained from outside sources acknowledged?

7. Body of Report. Read the entire body of the report, keeping in mind the following questions:
    - Does it present information to address the purpose of the report, and does it stay within the intended technical and geographical scope?
    - Are mathematical and chemical equations and formulas accurate, clear, numbered, referenced, and appropriate to the problem and methods used?
    - Does the text discuss the significance of the data in tables and illustrations and not just repeat the data?
    - Has written permission to use copyrighted material in the report been secured from the copyright holder?
    - Are the data shown in text, tables, and illustrations in agreement?
    - Has the discussion been developed along the main theme of the report as indicated in the title, table of contents, and purpose and scope?
    - Are all methods discussed relevant to the theme of the report? Do discussions answer the purpose of the report?
    - Is the report free of agency/company policy violations?

8. Summary, Conclusions, and/or Recommendations. The summary, conclusions, or summary and conclusions section is the terminal section of the report. A summary is a brief reaccounting of the informative parts of the report. The conclusions are answers to questions addressed by the purpose(s) of the report. The summary and conclusions are second in importance to the abstract or executive summary and usually serve as the principal source of information for those sections. The recommendations, if included, state what further action, if any, should be taken and the potential consequences of such actions.
    - Are the summary, conclusions, and/or recommendations a logical outgrowth of information report?
    - Does it contain or is it based only on information that is in the body of the report?
    - Does it reiterate the theme expressed in the title and purpose and scope?
    - Does it draw together and briefly reiterate the principal findings of the investigation?
    - Does it provide solutions or answers to problems addressed in the introduction?

- Is it as quantitative as possible and does it includes numerical findings presented in body of the report?

9. References. The list of references gives credit to the sources of all publications cited in the report.
   - What is the title of the list of references? If the list contains only references that are cited in the report, the list is titled *References* or *References Cited*. If the list is more extensive and contains references not cited in the report. If the list is an extensive or exhaustive compilation of pertinent references, it is called a *Bibliography*.
   - Are all literature citations in the text, tables, and illustrations listed? Has the author included the page, figure, or table number of the source material in the citation?
   - Are the references listed in the proper style and format for the intended publication?

10. Tables. Tables should be self-explanatory and in a format appropriate to the publication. See Section 8.3.2, "Editorial Review," for additional guidelines in reviewing tables.
    - Is the table needed? If so, are the data presented in a table unnecessarily repeated in the text? Are all data presented in the table needed?

11. Illustrations. Maps, hydrologic and geologic sections, graphs, diagrams, line drawings, and photographs should be self-explanatory. They should complement and support the text. They must be technically correct. Most problems with illustrations can be identified during a thorough editorial review.

12. Verification. Verification is the process that is intended to make the report internally consistent. Internal consistency can be improved by use of the following checklist:

    - Is the report title the same wherever it asppears—on the cover, title page, abstract or executive summary page, routing sheet, and transmittal memorandum?
    - Do values in the text, tables, and illustrations agree with one other?
    - Are the wording and level of headings in the table of contents the same as those in the body of the report?
    - Do figure and table titles agree with the lists in the table of contents?
    - Is the pagination correct?
    - Is the arithmetic in any calculations correct?
    - Are units of measurements in a consistent form and applicable, and are all included in the conversion table? Depending on the publication outlet, are all abbreviations that are used more than once spelled out in parentheses the first time they appear?
    - Are chemical, geologic, hydrologic, and other symbols in a standard format, and are they consistent throughout the report?

- Are all geographic names in the text, tables, and illustrations shown on a map or noted as being outside of map boundaries?
- Are contours shown on maps supported by values placed at data-control points on a review copy?
- Have changes made to the body of the report during review been incorporated in the abstract?

13. Reexamination. At this point, all parts of the report have been reviewed. Now go back and check it all over again. The reviewer has a good idea what the author has attempted to say, what the author really has said, and how the author has said it. A reexamination with all this in mind might disclose parts of the report where additional improvement is needed.

14. Review Memorandum. After the review has been completed, the reviewer should prepare a memorandum that summarizes the results of the review. Major problems should be described. Comments written in the manuscript need not be reiterated in the memorandum unless they have special significance. Comments of a complimentary nature also should be included in the memorandum.

ADDITIONAL RESOURCES

Association of Engineering Firms Practicing in the Geosciences (ASFE), 1992, *Important information about your geotechnical engineering report*. Silver Spring, MD: ASFE, 2p.

Bates, R. L., and Jackson, J. A., 1987, *Glossary of geology*, third edition. Alexandria, VA: American Geological Institute, 787 pp.

Cash, P., 1988, *How to develop and write a research paper*, second edition. New York: Prentice Hall, 96 pp.

CBE Style Manual Committee, CBE Style Manual, 1983, *A guide for authors, editors, and publishers' in biological sciences*. Bethesda, MD: Council of Biology Editors, Inc., 324 pp.

Day, R. A., 1988, *How to write and publish a scientific paper*, third edition. Phoenix: Oryx Press, 211 pp.

Heath, R. C., 1983, *Basic ground-water hydrology*, U.S. Geological Survey Water-Supply Paper 2220, 84 pp.

Hem, J. D., 1985, *Study and interpretation of the chemical characteristics of natural water*, third edition, U.S. Geological Survey Water-Supply Paper 2254, 263 pp.

Lamott, A., 1994, *Bird by bird, some instructions on writing and life*. New York: Anchor Books, 239 pp.

Lohman, S. W., 1972, *Ground-water hydraulics*, U.S. Geological Survey Professional Paper 708, 70 pp.

Lohman, S. W., et al., 1972, *Definitions of selected ground-water terms: revisions and conceptual refinements*, U.S. Geological Survey Water Supply Paper 1988, 21 pp.

Malde, H. E., 1986, *Guidelines for reviewers of geological manuscripts*. Alexandria, VA: American Geological Institute, 2 pp.

Moore, J. E., 1991, *A guide for preparing hydrologic and geologic projects and reports*. Dubuque, IA: Kendall/Hunt, 96 pp.

Moore, J. E., Aronson, D. A., Green, J. H., and Puente, C., 1988, *Report planning and review guide*, U.S. Geological Survey Open-File Report 88-320, 75 pp.

Moore, J. E., and Chase, E. B., 1982, *WRD project and report management guide*, U.S. Geological Survey Open-File Report 85-634, 243 pp.

Strunk, W. Jr., and White, E. B., 1979, *The elements of style*, New York: Macmillan, 85 pp.

U.S. Geological Survey, 1991, *Suggestions to authors of the reports of the United States Geological Survey*, seventh edition. Washington, D.C.: U.S. Government Printing Office, 289 pp.

U.S. Geological Survey, 1986, *Water Resources Division publications guide—Volume 1. Publications policy and text preparation*, U.S. Geological Survey Open-File Report 87-205, 429 pp.

U.S. Government Printing Office, 1984, *United States Government Printing Office style manual*. Washington, D.C.: Author, 479 pp.

Wilson, W. E., and Moore, J. E., 1998, *Glossary of hydrology*. Alexandria, VA: American Geological Institute, 248 pp.

9 Groundwater Development and Management

9.1 FEDERAL LAWS TO PROTECT GROUNDWATER

The federal government has passed strict, comprehensive, and long-term legislation that contains provisions for protecting groundwater and its quality. State governments have also passed similar, or even stricter, regulations.

The first law enacted to protect the nation's water resources is the Clean Water Act (CWA) of 1972 (33 U.S.C. §1251 et seq.). Although the law focuses mainly on protection of surface-water quality, it includes two provisions that have some effect on groundwater quantity and quality. Section 303 requires states to promulgate groundwater quality standards, and Section 208 requires that the designated state and local agencies develop comprehensive management plans for disposal of contaminants on land and in the subsurface to protect groundwater quality.

The principal law requiring that public water supplies are safe to drink is the Safe Drinking Water Act (SDWA) of 1974 (42 U.S.C §300f et seq.). The main objective of SDWA is the protection of groundwater drinking sources against contamination. The law established three programs: a system of national drinking-water standards; a system regulating the underground injection of wastes (UIC); and a sole source aquifer program (to protect aquifers that are the primary source of drinking water). The 1986 SDWA Amendments (P.L. 99-339) added significant new requirements to the three programs and a new program to protect areas subject to the influence of a well (wellhead protection areas).

The U.S. Congress amended the SDWA again in 1996 (P.L. 104-182). The amendments gave the U.S. Environmental Protection Agency (EPA) authority to target for regulation contaminants that could pose the greatest threat to public health and provided additional financial assistance for public water systems. The 1996 amendments include also right-to-know provisions, which require that the public must be notified—within 24 hours—of SWDA violations that present a threat to public health.

The Resource Conservation and Recovery Act (RCRA) of 1976 (42 U.S.C. §6901 et seq.) is designed to prevent waste disposal practices from threatening the environment (including groundwater). Key protections for groundwater include site-specific performance standards for treatment, storage, and disposal facilities; monitoring systems; and corrective action procedures. Additional requirements are included in the 1984 RCRA amendments.

The Comprehensive Environmental Response, Compensation, and Liability Act (CERCLA) of 1980 (42 U.S.C. §9601 et seq.)—also known as the "Superfund" law—authorizes EPA to initiate the protection of groundwater through removal and/or cleanup of hazardous substance disposal sites. The EPA may also seek compensation for the cost of cleanup and other corrective actions from responsible parties.

Marginally related to groundwater are two laws: the Federal Insecticide, Fungicide, and Rodenticide Act (FIFRA) as amended in 1988, which authorizes EPA to control pesticide use through registration; and the Toxic Substances Control Act (TSCA), under which EPA regulates a variety of chemicals.

In 1984, EPA issued a final draft of its Groundwater Protection Strategy (EPA, 1984). The main reasons for preparing this policy were (1) EPA authority to regulate potentially contaminating activities is fragmented among six statutes and (2) groundwater protection is not the main focus of any one statute. The strategy lays out approaches for achieving the following four goals:

1. Strengthen state groundwater protection programs.
2. Assess inadequately addressed groundwater problems (such as major unregulated sources of groundwater contamination).
3. Create a policy framework for guiding EPA programs.
4. Strengthen internal EPA groundwater organization.

The above-mentioned federal laws and the EPA Groundwater Protection Strategy provide a framework for action at the state level. Most of these statutes allow EPA to give states primary management responsibilities to carry out federal legislation.

The first attempt to summarize the state groundwater protection regulations and programs was done by the Committee on Groundwater Quality Protection of the National Research Council (NRC). The programs selected by the committee for review emphasize planning and regulatory aspects such as information gathering, classification systems, land use controls, and preventive enforcement systems. The committee classified groundwater protection program approaches into five categories:

1. Information collection and management systems
2. Classification systems
3. Groundwater quality standards
4. Control of contamination sources
5. Implementation of groundwater protection programs

First of all, a successful state program must be founded on an information base that allows proper definition of problems and evaluation of prevention strategies.

An important aspect of any regulatory program is the designation of lead agencies responsible for enforcing laws and regulations and for ensuring compliance with standards or ordinances. Enforcement of regulations and standards is perhaps the most difficult task of these agencies. Two alternative approaches can be used to encourage compliance: penalties (fines, taxes, loss of license) and incentives (tax credits, compensation, grants).

Today, state regulatory programs are well developed in each of the states. As with the federal Groundwater Protection Strategy, many states have or are moving toward adopting their own comprehensive strategies to protect groundwater supplies (e.g., Illinois, Minnesota, Nebraska, Wisconsin). EPA occasionally reports on the progress made by states in their groundwater protection programs.

9.2   TRANSBOUNDARY AQUIFERS

Transboundary aquifers cross a political boundary, and groundwater flows from one side of the boundary to the other. The aquifers might receive the majority of the recharge on one side while the majority of the water discharges on the other side of the political boundary.

These aquifers can only be defined by good observations and measurements of key hydrologic parameters. Even where international boundaries may follow rivers, the aquifers below may not reflect the same transfer from one side to another. The following transboundary aquifer systems are under study by national, regional, and international organizations (UNESCO, 2004):

- Guarani aquifer (South America)
- Nubian sandstone aquifers (Northern Africa)
- Karoo aquifers (Southern Africa)
- Slovak Karst-Aggtelek aquifer (Central Europe)
- Proded aquifer (Central Europe)
- Mexico/United States (El Paso, Texas)

A majority of the countries in the world share aquifers with neighboring countries. Political, socioeconomic, and other differences make the assessment and management of internationally shared aquifers difficult. Lack of knowledge of hydrogeology and specifically on the flow systems theory (Toth, 1998) can lead to poor understanding of groundwater flow and changes in water chemistry. An assessment of transboundary aquifers is needed to prevent or reduce groundwater problems.

Transboundary aquifers cover about 45 percent of the world's land area and affect 49 percent of the world's population. Eighty-nine transboundary aquifers have been identified in Europe.

9.3   EFFECT OF GROUNDWATER WITHDRAWAL

Groundwater in the United States is being increasingly developed for irrigation and municipal supplies. The U.S. Geological Survey (USGS) has noted that groundwater availability is a significant issue in almost every state (U.S. Geological Survey, 1984). The development of groundwater has led to declining water levels in a number of areas.

Under natural conditions, groundwater moves from areas of recharge to areas of discharge. The water is discharged to springs, lakes, streams, wetlands, or the ocean.

In most cases, equilibrium exists and long-term recharge of groundwater is balanced by long-term discharge.

Water levels in an aquifer fluctuate in response to changes in the rate of recharge and discharge. Overdraft occurs when discharge exceeds recharge. Water is released from storage and water levels fall. Groundwater overdraft (mining) occurs when there is the deliberate or inadvertent extraction of groundwater at a rate so in excess of replenishment that groundwater levels decline persistently, threatening exhaustion of the supply or at least a decline of pumping levels to uneconomic depths (Wilson and Moore, 1998).

Long-term withdrawals of water from a confined aquifer result in drainage of water both from clay layers within the aquifer and from adjacent confining beds. This drainage increases the load on the solid skeleton and causes compression of the aquifer and subsidence of the land surface.

Declines may amount to hundreds of meters in areas where large withdrawals from wells have caused discharge to exceed recharge over long periods of time. Six main areas in the United States where groundwater overdraft has had a major impact are Denver, Colorado; Texas High Plains; Tucson and Phoenix, Arizona; Chicago, Illinois; Central Valley California; and Las Vegas, Nevada.

Groundwater overdraft has caused environmental and other problems as follows:

- Stream-flow depletion
- Land subsidence
- Saltwater intrusion
- Increased cost to deepen wells
- Increased cost to pump groundwater
- Decrease in well yield
- Drying up of shallow wells and springs
- Degradation of water quality

Water conservation and regulations need to be implemented in areas where groundwater is being mined. Some measures that could be used to sustain the water supply are

- Artificial recharge
- Water reuse
- Conjunctive use of groundwater and surface water
- Restrictions on urban lawn irrigation and other water use
- Education at all levels on groundwater as a system

In some areas where groundwater is mined, land subsidence can occur when groundwater is withdrawn from a confined sand and gravel aquifer that contains highly compressible clays. As pressure in the aquifer drops, the clays become compacted and the land surface subsides. The consolidation results from a decrease in the pressure of water contained in the pores of the sediment. Consolidation in a groundwater aquifer system occurs in clay layers. The consolidation is not reversible.

TABLE 9.1
Major Areas of Land Subsidence

Locality	Subsidence		Area Affected Km 2
	Feet	m	
Central Valley, CA	30	9.0	13,500
Houston–Galveston, TX	9.1	2.75	12,170
Eloy, AZ	11.8	3.6	8,700
Tokyo	15.1	4.6	2,400
Po Valley, Italy	9.8	3.0	780
London	1.1	.35	450
Venice, Italy	.45	.14	400
Mexico City	28.5	8.70	225

The rock matrix and fluid pressure in the rock support the land surface. If clay and silt deposits are present, the surface may sink when water or oil is removed. Land subsidence results in a permanent reduction in the storage capacity of the confined aquifer. Substantial land subsidence from groundwater withdrawal has occurred in the San Joaquin Valley, California; Las Vegas, Nevada; Central Florida; Houston, Texas; Mexico City; Tokyo; Venice; and London. Table 9.1 lists major areas in the world where subsidence has occurred.

9.4   GROUNDWATER AND URBANIZATION

Urbanization is a major geomorphic process that affects both surface- and ground-water systems. The development of cities inevitably increases paved surfaces and roofs (termed *impervious cover*, which is not always impervious) that require storm drains. Installation of a network of other subsurface structures, including utility systems, is another necessary aspect of modern cities. Urbanization alters topography and natural vegetation, stream flows and flooding characteristics, temperatures both above and below the land surface, and water quality of surface streams and groundwater. Major physical changes to the groundwater system include changes in water table elevation, a dramatically altered permeability field created by construction and utility system emplacement, and altered groundwater recharge (Sharp, 2010).

ADDITIONAL RESOURCES

Alley, W. M., Riley, T. E. and Franke, O. L., 1999, *Sustainability of ground-water resources*, U.S. Geological Survey Water Supply Paper 1186, 79 pp.

EPA, 1985, *Practical guide to groundwater sampling*.

Moore, J. E., 2002, Effects of groundwater mining on aquifers—case histories: Denver, Tucson, Chicago, Central Valley, Las Vegas, IAH Congress Mar de Plata.

Sharp, J., 2010, Impacts of urbanization on groundwater systems-recharge, permeability, and geology, Geological Society of America (GSA) meeting, Denver, CO.

Tóth, J., 1998, Groundwater as a geological agent: an overview of the causes processes and manifestations, *Hydrogeology Journal*, vol. 7, no. 1, 11–14 pp.

U.S. Geological Survey, 1984, *National Water Summary 1984*, U.S. Geological Survey Water Supply Paper 2275, 467 pp.

U.S. Geological Survey, 1995, *Colorado, New Mexico, Nevada, and Arizona*, Hydrologic Atlas 730-C, 32 pp.

UNESCO, 2004, *Role of science*, Paris: Groundwater for Economic Development.

Wilson, W. E., and Moore, J. E., 1998, *Glossary of hydrology*. Alexandria, VA: American Geological Institute, 248 pp.

10 Case Studies

10.1 DENVER, COLORADO

The Denver Basin bedrock aquifer system is an important source of water for domestic, municipal, and agricultural uses in the Denver and Colorado Springs metropolitan areas (Figure 10.1). The Denver metropolitan area is one of the fastest growing areas in the United States, with a population of 1.2 million in 1960 that has increased with a corresponding increase in demand for potable water. Historically, the Denver area has traditionally relied on surface water; however, in the past 10 years new housing and recreation developments have begun to rely on groundwater from the bedrock aquifers as the surface water is fully appropriated and in short supply.

The Denver Basin bedrock aquifer system consists of Tertiary and Cretaceous age sedimentary rocks known as the Dawson, Denver, Arapahoe, and Laramie–Fox Hills aquifers. The number of bedrock wells has increased from 12,000 in 1985 to 33,700 in 2001, and the withdrawal of groundwater has caused water-level declines of 76 meters. Water-level declines for the past 10 years have ranged from 3 to 12 meters per year. The groundwater supplies were once anticipated to last 100 years, but there is concern that the groundwater supplies may be essentially depleted in 10 to 15 years in areas on the west side of the basin.

Extensive development of the aquifer system has occurred in the last 25 years, especially near the center of the basin in Douglas and El Paso Counties, where rapid urban growth continues and surface water is lacking. Groundwater is being mined from the aquifer system because the discharge by wells exceeds the rate of recharge. Concern is mounting that increased groundwater withdrawal will cause water-level declines, increased costs to withdraw groundwater, reduced well yield, and reduced groundwater storage. As the long-term sustainability of the groundwater resource is in doubt, water managers believe that the life of the Denver Basin aquifers can be extended with artificial recharge, water reuse, restrictions on lawn watering, well permit restrictions, and conservation measures.

10.2 MEXICO CITY

Mexico City is sinking at an alarming rate. Pumping of groundwater in Mexico City has resulted in water-level declines, obtained water quality changes, and severe land subsidence (Figure 10.2). As with many other cities in Mexico, Mexico City was established in a discharge area as the water table was shallow with water ready to be obtained; now all of them show subsidence effects. In Mexico City, the subsidence from 1932 to 1980 has been more than 8.7 meters. When the Spaniards conquered the Aztec capital in 1521, Mexico City straddled two large lakes, but the Spaniards drained the lakes. In early 1900, the annual sink in the city averaged 0.05 meters but

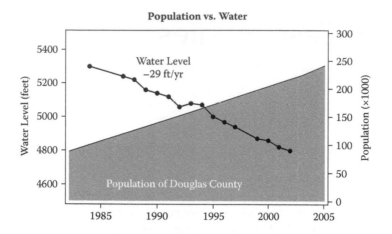

FIGURE 10.1   Denver population growth and water-level decline.

increased to 0.48 meters in the early 1950s. The springs stopped flowing early in the second half of the 20th century.

Groundwater use has increased because of industrial development and related population growth. Currently more than 24 million people live in Mexico City; in the early 1990s the city had an excess of 40 percent of the total industrial development of the country. The land has been subsiding at an average rate of 0.07 meters per year. The city started its subsiding process because water was withdrawn from the alluvial sand and gravel, and there was a water inducement from a saturated overlaying aquitard (5–10 percent sand-sized calcareous ooliths; 55–65 percent of silt-sized siliceous diatoms; 20–30 percent is clay-sized montmorillonitic; and 10 percent smectite and volcanic silica, bentonite (Diaz-Rodriguez and others, 1998); the remaining 5–10 percent is organic material and up to 10 wt percent organic carbon) (Figure 10.2). This layer with about 60 percent water content partially covers a set of aquifer units that have a lateral distribution far beyond the surface limits of the basin; they are hydraulically connected and have a thickness in excess of 3,000 meters.

Land subsidence, as it relates to groundwater, is a product of different factors that in practice would be difficult to measure separately at the field scale (Carrillo-Rivera et al., 2008). Since groundwater was originally pumped late in the 1800s, at least five responses (Bouwer, 1978) have resulted in observed land subsidence: (1) reduction of pore pressure due to local pumping, (2) replacement of cold water by warmer water (fluid pressure is a function of fluid temperature), (3) change of groundwater flow direction (regional upward groundwater hydraulic pressure effect), (4) increase of the load on the earth's surface (as city infrastructure), and (5) migration of saturation groundwater (from the aquitard unit). Although in theory each of these factors may be described in terms of the increase in effective stress (Freeze and Cherry, 1979) due to a reduction of fluid pressure, total stress = effective stress + fluid pressure. However, in practice

FIGURE 10.2   Mexico City church.

not all have been (and can be) measured, recognized, and monitored at the field scale. For instance, fluid pressure as a function of fluid temperature means that when the local cold groundwater flow obtained in pumping wells has been replaced beneath by warm water, the fluid stress has been reduced and consequently the effective stress has increased accordingly, producing a compaction in affected geological units that is manifested on the ground surface as land subsidence.

As groundwater pumping appears to be the main response to subsidence, it has been a general tendency to agree that the solution is to stop pumping; there are even some views that groundwater is required to be pumped at a rate less than current recharge. However, the processes below the Mexico City land surface need to be understood from a systemwide perspective. In other words, the migration of groundwater flow to wells needs to be clearly understood; it could be possible to have effective and natural subsidence controls that might prove to be more effective than an unfeasible relation of pumping to recharge, mainly when the aquifer limits are beyond the basin watershed.

In Mexico City groundwater pumping was estimated in late 1980s at >43 m/s, with about 70 percent of the pumping concentrated to the south of the watershed. Aquifer units (sediments and lava flows) have a collective thickness of more than 3,000 m, so regional flows have a contrasting temperature calculated (Edmunds, Carrillo-Rivera, and Cardona, 2002) at about 87°C for 1,800 meters' depth and a maximum of 162°C for the expected 3,000 meters' depth for regional flow. Boreholes initially constructed obtained cold water from local flows, but as extraction time has increased, regional flows are being induced at the well level. The temperature of obtained water has increased (Edmunds et al., 2002; Huizar-Alvarez and others, 2004). The response of thermal water inducement in land subsidence requires consideration for research as well as the other four responses above so as to understand the processes from a systemwide perspective.

Groundwater movement poses a driving force evident in the discharge area where an upward flow component prevails. A quicksand presence is related to a hydraulic head value larger than that of the weight of the sand grain. So when the force that is applied to lift a grain in a discharge area ceases due to a reduction in hydraulic head caused by regional groundwater pumping (i.e., away from the local pumping site), the fluid pressure is reduced, increasing the effective stress to maintain the value of total stress. The resulting increase in effective stress (grain compaction) is manifested in the vertical scale as land subsidence. The original groundwater condition of the Mexico City plane was that of a discharge area; reports of artesian pressure of more than 30 meters was not uncommon. The lowering of this pressure is due to (1) uncontrolled additional pumping beyond the limits of the city, (2) distortion to the hydraulic head related to the deviation of flow to another area, and (3) production of new recharge conditions with vertical flow downward, all of which involve secondary effects of subsidence enhancement.

The increase of the load on the earth's surface needs more consideration; most urban efforts have started with the construction of required infrastructure at levels that did not considerably and visually disturb the soil. However, as the city has grown, the population has also increased and the volume and weight of the related infrastructure has also been augmented enormously. From this perspective, six main parameters need to be considered related to subsidence at a local scale: compressibility of underlying geological materials, their thickness, reduction in fluid pressure, measurement of ground-surface elevation, groundwater elevation, and applied infrastructure weight.

Migration of groundwater is a process that is considered paramount in the case of land subsidence in Mexico City; it occurs when groundwater is extracted from an aquifer unit located beneath the aquitard covering almost the entire surface of the city. The aquitard releases a considerable volume of its stored water (up to 60 percent in volume) due to the difference in hydraulic gradient created by pumping in the aquifer unit beneath. Extraction diminishes the hydraulic head in the groundwater of the aquifer unit, permitting aquitard water to flow downward due to the new prevailing hydraulic gradient. The migration of this groundwater takes place through the surface of contact between the aquitard–aquifer unit; although the vertical hydraulic conductivity of the aquitard is low (around 6×10^{-9} ms^{-1}; permitting the flow to occur at the expense of the water stored in the aquitard. The volume of migration groundwater is large at the regional scale, and as the aquitard has no replenishment due to its low hydraulic conductivity, this flow is made at the expense of aquitard stored water. The loss of migrated water in the aquitard is consequently manifested as a reduction in aquitard volume that is manifested at the surface as land subsidence.

This migration of water is the component given traditional consideration in land subsidence response reported for Mexico City in numerous studies dating from 1966. The leaky or semiconfined aquifer test conditions were in Mexico City the daily expected response to aquifer-test interpretation of any "s-t" curve, so the Hantush (1956) type curve method was currently applied in Mexico and elsewhere considering that a semiconfined case was to be solved. However, from this perspective of migration of groundwater, Huizar-Alvarez and others (2004) demonstrated that the semiconfined condition in the "s-t" field curve response is not the rule anymore, as in southern Mexico City. Field evidence shows hydraulic discontinuity between the aquitard and

FIGURE 10.3 Nubian aquifer flowing well.

the granular aquifer; the s-t semi-confined response is manifested by regional flow beneath.

This reduction of flow from the aquitard to the aquifer unit beneath seems to be a plausible conclusion for the reduction of land subsidence in the Xochimilco zone. Values for 1980s were about −0.49 to −0.25 meters annum[-1], and in the 1990s land subsidence velocity was lower at −0.28 to −0.22 meters annum[-1]; this tendency continued and for year 2000 reported values ranged from −0.18 meters annum[-1] to positive values of 0.02 meters annum[-1] (Carrillo-Rivera and others, 2002). Such values appear to be supported as current extraction has lowered hydraulic heads some 46 meters below the aquitard floor in the Xochimilco zone (Angeles-Serrano and others, 2008).

It can be deduced from above that land subsidence is related to geologic evolution and to properties and distribution of sediments, rocks, and fluids. It is also paramount to consider that groundwater extraction has a different effect on the groundwater flow regime if it is tapped in a recharge, a transit, or a discharge area. Further, the type of flow hierarchy (local, intermediate, or regional) and the particular site characteristics would be impacted considerably in observed response of land subsidence.

10.3 NUBIAN AQUIFER, NORTHERN AFRICA

One of the world's most extensive aquifers is the Nubian sandstone, which underlies Egypt, Chad, Libya, and Sudan in the Sahara desert (Figure 10.3). The aquifer is composed of hard ferruginous sandstone with clay layers with a thickness of 140–230 meters. Groundwater salinity ranges from 240–1,300 mgl. The area is arid and population is heavily dependent on the aquifer for domestic, industrial, and agricultural supply. The aquifer recharge is negligible, and since it is much less than the withdrawal from groundwater, it is considered to be nonrenewable.

The aquifer system is not in equilibrium. The groundwater flow is from north to south and the discharge is from natural evaporative areas. The Nubian aquifer has been used as a water supply in the desert areas for millennia. Libya has an ambitious program (Great Man-Made river project) of pumping from the Nubian, which has resulted in water-level declines and saltwater intrusion. Egypt is developing resort sites in the western and eastern deserts with Nubian water.

10.4   CALIFORNIA'S CENTRAL VALLEY

The Central Valley of California is one of the most intensively developed areas of irrigated agriculture (Figure 10.4). The valley is 800 kilometers miles long by 80 kilometers wide. The valley is underlain by a large confined alluvial aquifer to a depth of 7,600 meters. Fresh groundwater is present to depths of as much as 1,220 meters, but most wells are less than 300 meters. Significant development of groundwater for irrigation began in the early 1900s. As groundwater withdrawal increased to the point that groundwater withdrawal exceeded recharge, the surface started to decline. In some areas, potentiometric surface declined 79 meters (1940 and 1963). The groundwater withdrawal caused the compaction of fine-grained confining beds, which resulted in subsidence of the land surface. Water level declines of 48 meters or more have resulted in land subsidence of about 10 meters in the Los Banos–Kettleman City area.

10.5   CHICAGO, ILLINOIS

Since 1864 large quantities of groundwater have been withdrawn from glacial drift and bedrock aquifers for municipal and industrial use. In the Chicago area, two Cambrian-Ordovician bedrock aquifers (sandstone and dolomite) supply most of the water. One of the first wells in the area was drilled to a depth of 216 meters and flowed at the land

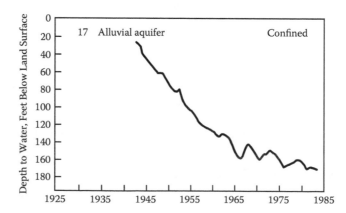

FIGURE 10.4   Central Valley, California. (From Moore, J., *Field Hydrology: A Guide for Site Investigations and Report Preparation*, CRC Press, Boca Raton, 2002. With permission from Taylor & Francis.)

surface /s-1 (Alley and others, 1999). Groundwater withdrawal from the Cambrian-Ordovician bedrock aquifers has caused water-level (artesian head) declines of more than 244 meters. No major land subsidence has been reported as a result of large withdrawals because the rocks in the area are consolidated and resist compaction. A principal concern has been the potential converts the confined aquifer to an unconfined aquifer and thus allows mining to take place. Since 1980 many public water supplies were shifted to water from Lake Michigan. The shift has resulted in a significant decrease in withdrawal and general recovery of potentiometer heads. Pumping continues in all parts of the area, however, and may be increasing in some areas.

10.6   LAS VEGAS, NEVADA

The Las Vegas valley is one the fastest growing metropolitan areas in the United States. Water to support this rapid growth is being supplied by imported Colorado River water and local groundwater. Groundwater is pumped from the upper 600 meters of unconsolidated alluvial sediments. The deeper aquifers, called "principal artesian aquifers" (below 300 feet) are capable of yielding large quantities of groundwater. Overlying the principal aquifers are 30- to 100-meter-thick deposits of clay. Since the 1970s the annual groundwater withdrawal has remained between 60,000 and 90,000-acre feet.

Most of the withdrawal is from the northwest area, which has seen water-level declines of more than 100 meters. Areas in central Las Vegas (the strip) have experienced declines of as much as 60 meters. Since 1935 compaction of the aquifer has caused nearly 1.5 meters (6 feet) of subsidence and has led to the formation of numerous earth fissures and surface faults. Before groundwater development, the aquifer sustained the flow of many springs that discharged into the Las Vegas Wash (*las vegas* means "the springs"), but groundwater withdrawal has caused these springs to dry up. Urban runoff has created a reservoir of poorer quality; this contaminated water now recharges the principal artesian aquifer.

10.7   TUCSON AND PHOENIX, ARIZONA

Large water-level declines in the unconfined alluvial aquifers have caused land subsidence and earth fissures to develop in a 3,000-square-mile area that includes Tucson and Phoenix. Water levels in the Tucson area have declined 49 meters from 1945 to 1984. The subsidence is caused by compaction of fine-grained sediments in the basin fill. The sediments deform and compact when water-level declines subject the sediments to additional compression from the weight of the overlying deposits. Compaction increases slowly as the water levels decline. Compaction and land subsidence has caused cracks to develop in the land surface, which can extend for hundreds to thousands of meters along the surface and can be hundreds of meters deep.

10.8   HIGH PLAINS AQUIFER, SOUTHWEST UNITED STATES

The High Plains aquifer underlies one of the major agricultural areas in the world. The aquifer includes parts of eight states: Colorado, Kansas, Nebraska, New Mexico, Oklahoma, South Dakota, Texas, and Wyoming.

The area is the remainder of a vast plain formed by alluvial sediments by streams flowing from the ancestral Rocky Mountains. The saturated thickness of the aquifer ranges from zero to 300 meters and averages 60 meters. The mean annual precipitation ranges from 0.40 to 0.71 meters. Evaporation rates are high compared to precipitation, and thus little water is available to recharge the aquifer.

Farmers began extensive use of groundwater for irrigation starting in the 1930s and 1940s. Groundwater has declined during the past 50 years. Decline areas range from 3 to more than 30 meters. In 1949 about 326 million cubic per day of groundwater was used for irrigation and in 1980 about 704 million cubic /s. The southern High Plains is perhaps one of the best known examples of long-term declines of water level due to lack of equilibrium between recharge and discharge.

10.9   BANGKOK, THAILAND

Thailand originally obtained its water supply from the Chao River. However, in 1950 the river became so polluted that water was obtained from the alluvial aquifer beneath the city. The pumpage caused subsidence of 10 cm per annum in the southern and eastern parts of the city. From 1987 to 2003 the subsidence totaled 1 meter.

10.10   TOKYO, JAPAN

Overexploitation of groundwater in Tokyo with a population of 12.4 million has led to subsidence and sea water intrusion. These problems have been resolved by better management of groundwater use. The subsidence in Tokyo reached 15.1 feet in the 1970s. In 1989 groundwater quality management was formulated. Subsidence for the most part is under control because surface water is imported for industrial use.

ADDITIONAL RESOURCES

Alley, W. M., Riley, T. E., and Franke, O. L., 1999, *Sustainability of ground-water resources*, U.S. Geological Survey Circular 1186, 79 pp.

Bouwer, H., 1978, *Groundwater hydrology*, Water Resources and Environmental Engineering Series. Sydney: McGraw-Hill, 480 pp.

Ortega-Guerrero and others, 1993, The (saturated) hydraulic conductivity of the aquifer unit is higher, in the range of 1.5×10^{-3} to 8.1×10^{-8} m s^{-1} (Vázquez-Sanchez, 1995)

Díaz-Rodríguez, A. et al., 1998, Physical, chemical and mineralogical properties of Mexico City sediments: a geotechnical perspective. *Canadian Geotechnical Journal*, vol. 35, 600–610 pp.

Edmunds, W. M., Carrillo-Rivera, J.J., and Cardona, A., 2002, Geochemical evolution of groundwater beneath Mexico City. *Journal of Hydrology*, vol. 258, 1–24 pp.

Huizar-Alvarez, R., and others, 2004, Chemical response to groundwater extraction southeast of México City, *Hydrogeology Journal*, vol. 12, 436–450 pp.

Fetter, C. W., 1994, *Applied hydrogeology*. Macmillan.

Winter, T. C., 1999, *Ground water and surface water, a single resource*, U.S. Geological Survey Circular 1139, 76 pp.

Weight, W. D. 2008, *Hydrogeology field manual*. McGraw-Hill.

Glossary

Abandoned well: A well whose use has been permanently discontinued or which is in a state of such disrepair that it cannot be used for its intended purpose (U.S. Environmental Protection Agency, 1994).

Absorption: The uptake of water, other fluids, or dissolved chemicals by a cell or an organism (U.S. Environmental Protection Agency, 2001).

Acre foot (AF): The quantity of water required to cover 1 acre to a depth of 1 foot and is equal to 43,560 cubic feet or about 326,000 gallons or 1,233 cubic meters.

Affluent stream: Stream flowing toward or into a larger stream or lake.

Agent Orange: A toxic herbicide and defoliant used in the Vietnam conflict (U.S. Environmental Protection Agency, 2001).

Aggregate: Hard materials such as sand, gravel, and crushed stone used for mixing with cementing material to form concrete, mortar, or asphalt.

Alluvium: A general term for all detrital deposits resulting directly or indirectly from the sediment transport of streams, thus including the sediments laid down in riverbeds, flood plains, lakes, fans, and estuaries. A general term for clay, silt, sand, gravel, or similar unconsolidated detrital material deposited during comparatively recent geologic time by a stream or other body of running water, as sorted or semisorted sediment in the bed of the stream or on its floodplain or delta, as a cone or fan at the base of a mountain slope (AGE, 1987).

Alkaline: The condition of water or soil that contains a sufficient amount of alkali substance to raise the pH above 7.0 (U.S. Environmental Protection Agency, 2001).

Aquiclude: A saturated geologic unit that is incapable of transmitting significant quantities of water under ordinary hydraulic gradients (Freeze and Cherry, 1979). Replaced by the term *confining bed* (Lohman and others, 1972, p. 2).

Aquifer: A formation, group of formations, or part of a formation that contains water to wells and springs (Lohman and others, 1972, p. 2). Also called *water-bearing formation*.

Aquifer system: A heterogeneous body of permeable material that acts as a water-yielding hydraulic unit of regional extent.

Aquifer test: A test involving the withdrawal of measured quantities of water from, or addition of water to, a well and the measurement of resulting changes in hydraulic head in the aquifer both during and after the period of discharge or addition.

Aquitard: A saturated unit of low hydraulic conductivity that can store and slowly transmit groundwater (Weight, 2008).

Area of influence: The area surrounding a pumping or recharging well within which the potentiometric surface has been changed.

Artesian aquifer: See *confined aquifer.*

Artesian head: The hydrostatic head of an artesian aquifer or of the water in the aquifer (AGE, 1980, p. 36).

Artesian pressure: Hydrostatic pressure of artesian water, often expressed in terms of pounds per square inch at the land surface, or in terms of height in feet above the land surface, of a column of water that would be supported by the pressure (AGE, 1980, p. 36).

Artesian spring: A spring from which the water flows under artesian pressure, usually through a fissure or other opening in the confining bed above the aquifer (AGE, 1980, p. 36).

Artesian well: A well tapping confined groundwater. Water in the well rises above the level of the water table under artesian pressure, but does not necessarily reach the land surface. Sometimes restricted to mean only a flowing artesian well (AGE, 1980, p. 36).

Backwashing: A process of well development. In this process water in the well casing is lifted by air for several minutes followed by shutting off the air for several minutes. This action is repeated several times until the water is clear.

Bailer: A pipe with a valve at the lower end, used to remove slurry from the bottom or side of the well as it is drilled (U.S. Environmental Protection Agency, 2001).

Baseflow: That part of stream discharge that is groundwater seeping into the stream (Fetter, 1994, p. 636). That part of the stream discharge that is not attributable to direct runoff from precipitation or melting snow. It is sustained by groundwater discharge (Groundwater Subcommittee, 1989).

Basic hydrologic data: Includes inventories of features of land and water that vary only from place to place (topographic and geologic maps are examples) and records of processes that vary with both place and time (Langbein and Iseri, 1960).

Bladder pump: A positive-displacement pumping device that uses pulses of gas to push a water-quality sample toward the surface (Fetter, 1994).

Baseflow recession hydrograph: A hydrograph that shows a baseflow-recession curve (Fetter, 2001).

Bedrock: A general term for the rock, usually solid, that underlies soil or other unconsolidated, superficial material. (AGE, 1980, p. 60). The solid rock underlying soils and regolith or exposed at surface (ASTM, 1994).

Bentonite: A colloidal clay expandable when moist, commonly used to form a tight seal around well casing (U.S. Environmental Protection Agency, 2001).

Beneficial use: A use of water resulting in appreciable gains or benefits to the user consistent with state law (Vandas, Winter, & Battaglin, 2002).

Biologic oxygen demand (BOD): An index of the amount of oxygen consumed by living organisms (mainly bacteria) while utilizing the organic matter in waste (Dunne and Leopold, 1978).

Bioremediation: Use of microorganisms to control and destroy contaminants (National Academy, 1993).

Borehole: A circular hole in the ground made by boring. Commonly, a deep hole of small diameter, as an oil well or a water well (AGE, 1980, p. 76).

Borehole geophysics: The application of certain physical principles, such as magnetic attraction, gravitational pull, speed of sound waves, and the behavior of electric currents, to determine the nature of rock and water penetrated by a borehole A general field of geophysics developed around the lowering of various probes into a well (Fetter, 1994).

Boring: A hole advanced into the ground by means of a drilling rig (Fetter, 1994).

Brackish water: (1) An indefinite term for water with a salinity intermediate between that of normal seawater and that of normal freshwater (AGE, 1980, p. 79). (2) Water that contains more than 1,500 mg/L but not more than 15,000 mg/L total dissolved solids.

Cable tool drilling: A method of drilling. Rock at the bottom of the hole is broken up by a steel bit with a blunt, chisel-shaped cutting edge. The bit is at the bottom of a heavy string of steel tools suspended on a cable that is activated by a walking beam, the bit chipping the rock by regularly repeated blows. The method is adapted to drilling water wells and relatively shallow oil wells (AGE, 1980, p. 87).

Calibration: The process of fitting a model to a set of observed data by changing unknown or uncertain model parameters systematically within their allowable ranges until a best fit of model to the observed data is achieved (Fetter, 1994).

Capillary fringe: The zone above the water table within which the porous medium is saturated by water under less than atmospheric pressure (U.S. Environmental Protection Agency, 2001).

Cation exchange: The exchange of cations between a solution and those cations held on the outer surface of mineral or organic matter in the soil (Loynachan and others, 1999).

Cistern: Small tank or storage facility used to store water for a home or farm; often used to store rainwater (U.S. Environmental Protection Agency, 2001).

Coal bed methane: Methane absorbed onto the surface of coal and which may be produced when pressure conditions are reduced to allow the gas to be released (Harrison and Testa, 2003).

Cone of depression: A depression of the potentiometric surface in the shape of an inverted cone that develops around a well that is being pumped.

Conductivity: The ability of a material to conduct electrical current. Units are siemans per meter (Weight, 2008).

Confined aquifer: An aquifer that is overlain by a confining bed (Fetter, 1994, p. 635). *Artesian aquifer.*

Confined groundwater: Groundwater under pressure significantly greater than that of the atmosphere. Its upper surface is the bottom of an impermeable bed or a bed of distinctly lower permeability than the material in which the water occurs (AGE, 1980, p. 132). *Artesian water.*

Confined-water well: A well for which the sole source of supply is confined groundwater (Rogers and others, 1981, p. 75). *Artesian well.*

Confining layer: A layer (bed) of impermeable material stratigraphically adjacent to one or more aquifers. In nature, its hydraulic conductivity may range from nearly zero to some value distinctly lower than that of the aquifer. It supplants aquifuge, aquitard, and aquiclude (After Lohman et al., 1972, p. 5). A body of material of low hydraulic conductivity that is stratigraphically adjacent to one or more aquifers. It may be above or below the aquifer (Fetter, 1994).

Conjunctive use of water: The joining together of two sources of water, such as groundwater and surface water, to serve a particular use.

Connate water: Water entrapped in the interstices of a sediment at the time of its deposition; recommended that it be defined as "water that has been out of contact with the atmosphere for at least an appreciable part of a geologic period." Commonly misused by reservoir engineers and well-log analysts to mean interstitial water or formation water (AGE, 1980, p. 133). *Fossil water.*

Consolidation: Any or all of the processes whereby loose, soft, or liquid earth materials become firm (Bucksch, 1996).

Constant head boundary: The conceptual representation of a natural feature such as a lake or river that effectively fully penetrates the aquifer at that location (ASTM, 1994).

Consumptive use: A term used mainly by irrigation engineers to mean the amount of water used in crop growth plus evaporation from the soil (USSCS, 1971). Water absorbed by the crop and transpired or used directly in the building of plant tissue together with that evaporated from the cropped area (U.S. Army Corps of Engineers, 1991).

Contact spring: A type of gravity spring whose water flows to the land surface from permeable strata over less permeable or impermeable strata that prevent or retard the downward percolation of the water (Meinzer, 1923, p. 51).

Contaminant: Any physical, chemical, biological, or radiological substance or matter that has an adverse affect on air, water, or soil (U.S. Environmental Protection Agency, 1994). *Pollutant.*

Contamination: As applied to water, the addition of any substance or property preventing the use or reducing the usability of the water for ordinary purposes such as drinking, preparing food, bathing, washing, recreation, and cooling. Sometimes arbitrarily defined differently from pollution, but generally considered synonymous.

Core: In drilling, a cylindrical column of rock taken as a sample of the interval penetrated by a core bit and brought to the surface for geologic examination and/or laboratory analysis (AGE, 1980, p 140). *Drill core.*

Core log: A record showing, for example, the depth, character, lithology, porosity, permeability, and fluid content of a core (AGE, 1980, p. 141).

Crest gauge: A device for obtaining the elevation of the flood crest of streams or lakes (Rantz and others, 1982, p. 77). *Crest-state gauge.*

Crystalline rock: An inexact but convenient term designating an igneous or metamorphic rock as opposed to a sedimentary rock (AGE, 1987). An igneous or metamorphic rock consisting wholly of relatively large mineral grains.

Cubic feet per second (CFS): A unit expressing rates of discharge (Langbein and Iseri, 1960).

Current: (1) The flowing of water or other liquid. (2) That portion of a stream of water that is moving with a velocity much greater than the average or in which the progress of the water is principally concentrated (Rogers et al., 1981, p. 87).

Current meter: Instrument for measuring the speed of flowing water (Lohman et al., 1960, p. 7). The U.S. Geological Survey uses a rotating cup meter.

Daily hydrograph: A hydrograph showing days as the unit of time.

Data: Records of observations or measurements of facts, occurrences, and conditions in written, graphical, or tabular form (U.S. Army Corps of Engineers, 1991).

Darcy's Law: An empirical law that states that the velocity of flow in a permeable media is directly proportional to the hydraulic gradient assuming that the flow is laminar and inertia can be neglected.

Deep well: Water wells, generally drilled, extending to a depth greater than that typical of shallow wells in the vicinity. The term may be applied to a well 20 m deep in an area where shallow wells average 7 or 8 m deep, or to a much deeper well in an area where the shallowest aquifer supplies wells 100 m deep or more (AGE, 1980, p. 163).

Delayed yield: The second or delayed response after a recharge response indicated on the field plot of time drawdown data during an aquifer test (Weight, 2008).

Dense nonaqueous phase liquid (DNAPL): A polluting liquid with a density greater than water that sinks to the base of the aquifer, for example, creosote and trichloro-ethylene (Stanger, 1994). Denser-than-water nonaqueous phase liquid synonymous with denser-than-water immiscible phase liquid (Cohen and Mercer, 1993).

Discharge area: An area in which there are vertical components of hydraulic head in which head increases downward the aquifer. Groundwater is flowing toward the surface in a discharge area and may exit as a spring, seep, or stream baseflow or by evaporation and transpiration.

Dispersion: The phenomenon by which solute in flowing groundwater is mixed with uncontaminated water and becomes reduced in concentration. Dispersion is caused by both differences in the velocity that the water travels at the pore level and differences in the rate at which water travels through different strata in the flow path (Fetter, 1994, p. 639).

Distribution coefficient: The quantity of the solute sorbed by the solid per unit weight of solid divided by the quantity dissolved in the water per unit volume of water (Cohen and Mercer, 1993).

Distilled water: Water formed by the condensation of steam or water vapor (Rogers et al., 1981, p. 107).

Domestic water: Water used in homes and on lawns, including use for laundry, washing cars, cooling, and swimming pools (Wang, 1974, p. 123).

Double porosity: Water can be withdrawn from both the fractures and matrix of a fractured porous medium and discrete flow in individual fractures (Dominico and Schwartz, 1990).

Drawdown: The lowering of the surface elevation of a body of water, the water surface of a well, the water table, or the potentiometric surface adjacent to the well, resulting from the withdrawal of water. The difference between the nonpumping water level at some time and pumping level at that time (U.S. Army Corps of Engineers, 1991). The lowering of the water table of an unconfined aquifer or the potentiometric surface of a confined aquifer caused by pumping of groundwater from wells (Fetter, 1994, p. 238).

Dug well: Well excavated by means of picks, shovels, or other hand tools or by means of a power shovel or other dredging or trenching machinery, as distinguished from one excavated by a drill or auger.

Effective porosity: The amount of interconnected pore space available for fluid transmission. It is expressed as a percentage of the total rock volume that is occupied by the interconnecting interstices. Although *effective porosity* has been used to mean about the same thing as *specific yield*, such use is discouraged (Lohman et al., 1972). This definition of *effective porosity* differs from that of Meinzer (1923, p. 28).

Effluent: Wastewater, treated or untreated, that discharges from a factory, sewer works, or treatment plant (Vandas et al., 2002).

Electrical well log: A record obtained in a well investigation in rock from a traveling electrode; it is in the form of curves that represent the apparent values of the electric potential and electric resistivity or impedance of the rocks and their contained fluids throughout the uncased portions of a well (Rogers et al., 1981, p. 1260).

Environmental Impact Statement (EIS): A legally required document to describe the environmental impact of a project (Stanger, 1994).

Ephemeral lake: Short-lived lake (AGE, 1980, p. 206).

Ephemeral stream: A stream or reach of a stream that flows briefly only in direct response to precipitation in the immediate locality and whose channel is at all times above the water table. The term may be arbitrarily restricted to a stream that does not flow continuously during periods of as much as one month (Meinzer, 1923, p. 58).

Equipotential line: A line, in a field of flow, such that the total head is the same for all points on the line, and therefore the direction of flow is perpendicular to the line at all points (U.S. Army Corps of Engineers, 1991).

Equipotential surface: A surface in a three-dimensional groundwater flow field such that the total hydraulic head is the same for all points along the line (Fetter, 1994).

Eutrophic lakes: Murky bodies of water with concentrations of plant nutrients causing excessive production of algae (U.S. Environmental Protection Agency, 2001).

Evaporation: The process by which water passes from the liquid to the vapor state (Fetter, 1980). The quantity of water that is evaporated; the rate is expressed as depth of liquid water removed from a specified surface per unit of time, generally in inches or centimeters per day, month, or year. The concentration of dissolved solids by driving off water through the application of heat (Rogers et al., 1981, p. 132).

Evapotranspiration: Water withdrawn from the land area by evaporation from the water surfaces and moist soil and plant transpiration (Langbein and Iseri, 1960).

Expansive soil: Soil that expands or shrinks as its moisture content changes (Holzer, 2009).

Facies change: A lateral or vertical variation in the lithologic character of sedimentary deposits.

Fault spring: Spring flowing onto the land surface from a fault whose movement resulted in a permeable bed coming in contact with an impermeable bed (AGE, 1980, p. 224).

Fecal coliform bacteria: Bacteria found in the internal tracts of mammals. Their presence in water is an indicator of pollution and possible contamination by pathogens (U.S. Environmental Protection Agency, 2001).

Field capacity: The quantity of water that can be permanently retained in the soil in opposition to the downward pull of gravity (Langbein and Iseri, 1960, p. 9). *Field moisture capacity.*

Finite-difference model: A digital groundwater flow model based upon a rectangular grid that sets the boundaries of the model and nodes where the model will be solved (After Fetter, 1994).

Finite-element model: A digital groundwater flow model where the aquifer is divided into a mesh formed by multiple polygonal cells (After Fetter, 1994).

Flood plain: The surface or strip of relatively smooth land adjacent to a river channel constructed by the present river in its existing regimen and covered with water when the river overflows its bank (Vandas et al., 2002).

Flowing well: A well that discharges water at the surface without the aid of a pump or other lifting device; a type of artesian well (Rogers et al., 1981, p. 153).

Formation water: Water present in a water-bearing formation under natural conditions (Harrison and Testa, 2003).

Fracture: A joint or fault in rock or rock material.

Fracture permeability: The property of a rock or rock material that permits movement of water along interconnecting joints and faults.

Fracture spring: Spring whose water flows from joints or other fractures (AGE, 1980, p. 244).

French drain: An underground passageway for water through the interstices among stones placed loosely in a trench (U. S. Army Corps of Engineers, 1991).

Freshwater: Water that contains less than 1,000 mg/L of dissolved solids; generally more than 500 mg/L is undesirable for drinking and many industrial uses (National Academy, 1993).

Fringe water: Water of the capillary fringe.

Fully penetrating well: Well that penetrates and is open to entire saturated thickness of the aquifer.

Gauge: A device for indicating the magnitude or position of an element in specific units when such magnitude or position is subject to change; examples of such elements are the elevation of a water surface, the velocity of flowing water, the pressure of water, the amount or intensity of precipitation, and the depth of snowfall (After Rogers et al., 1981, p. 162).

Gauge height: The water-surface elevation referred to some arbitrary gauge datum. *Gauge height* is often used interchangeably with the more general term *stage*, although *gauge height* is more appropriate when used with a reading on a gauge (Langbein and Iseri, 1960, p. 11).

Gauging station: A site on a stream, canal, lake, or reservoir where systematic observations of gauge height or discharge are obtained (Langbein and Iseri, 1960, p. 11).

Gaining stream: A stream or stretch of stream that receives water from groundwater in the saturation zone. The water surface of such a stream stands at a lower level than the water table or piezometric surface of the groundwater body from which it receives water. *Effluent stream* (Rogers et al., 1981, p. 163).

Geographic information system (GIS): An organized collection of computer hardware, software, geographic data, and personnel designed to efficiently collect, store, update, manipulate, analyze, and display all forms of geographically referenced information.

Geohydrology: The study of earth science aspects of groundwater. Emphasis is usually on the hydraulic groundwater flow in aquifers. Also used in reference to all hydrology on the Earth without restriction to geologic aspects (Stringfield, 1966, p. 3).

Geologic map: A graphic representation of geologic information, such as the distribution, nature, and age relationships of rock units (surficial deposits may or may not be mapped separately), and the occurrence of structural features (folds, faults, joints), mineral deposits, and fossil localities. It may indicate geologic structure by means of formational outcrop patterns, by conventional symbols giving the direction and amount of dip at certain points, or by structure-contour lines (AGE, 1980, p. 258).

Ganats: Long, gently sloping tunnels used to collect groundwater in North Africa, Egypt, and Iran to collect groundwater. Because the water flows under gravity, the tunnels do not have to be pumped.

Glacial aquifer: Material deposited by a glacier, or in connection with glacial processes, that is capable of yielding water to wells.

Gravel: Unconsolidated coarse, granular material, larger than sand grains, resulting from reduction of rock by natural or artificial means. Sizes range from 0.16 cm (no. 4 sieve) to 2.6 cm in diameter. Coarse gravel ranges from 7.6 cm to 1.9 cm, whereas fine gravel ranges from 1.9 cm to 0.16 cm.

Gravity drainage: Refers to the movement of DNAPL in an aquifer that results from the force of gravity (Cohen and Mercer, 1993).

Gravity flow: Movement of glacier ice as a result of the inclination of the slope on which the glacier rests (AGE, 1980, p. 274).

Gravity water: A supply of water that is transported from its sources to its place of use by means of gravity, as distinguished from a supply that is pumped (Rogers et al., 1981, p. 169).

Ground improvement: Engineering techniques that stabilize problem soils, decreasing their compressibility and increasing their strength (Holzer, 2009)

Groundwater: Water in the ground that is in the zone of saturation, from which wells, springs, and groundwater runoff are supplied (Langbein and Iseri, 1960). The part of the subsurface water that is in the saturated zone. Loosely, all subsurface water (excluding internal water) as distinct from surface water (After AGE, 1980, p. 277). Water in the saturated zone beneath the water table (USSCS, 1971).

Groundwater barrier: A natural or artificial obstacle, such as a dike or fault gouge, to the lateral movement of groundwater, not in the sense of a confining bed.

It is characterized by a marked difference in the level of the groundwater on opposite sides (AGE, 1980, p. 278). *Groundwater dam.*

Groundwater basin: (1) A subsurface geologic structure having the character of a basin with respect to the collection, retention, and outflow of water. (2) An aquifer or system of aquifers, whether basin shaped or not, that has reasonably well-defined boundaries and more or less definite areas of recharge and discharge (AGE, 1980, p. 278).

Groundwater budget: A numerical account, the groundwater equation, relating recharge, discharge, and changes in storage of an aquifer, part of an aquifer, or system of aquifers (AGE, 1980, p. 278). *Groundwater inventory.*

Groundwater contamination: *Groundwater pollution.*

Groundwater depletion: The lowering of the groundwater level of an area (Rogers et al., 1981, p. 174). The decrease in available groundwater supplies.

Groundwater flow: See *groundwater runoff* and *groundwater movement.*

Groundwater irrigation: Irrigation by water derived from wells, springs, or shallow ponds not supplied by surface runoff (Rogers and others, 1981, p. 173).

Groundwater level: The elevation of the water table or other potentiometric surface at a particular place or in a particular area, as represented by the level of water in wells or other natural or artificial openings or depressions communicating with the zone of saturation (After AGE, 1980, p. 278).

Groundwater movement: The flow of groundwater in response to hydraulic gradient.

Groundwater outflow: That part of the discharge from a drainage basin that occurs through the aquifer. The term *underflow* is often used to describe the groundwater outflow that takes place in alluvium and thus is not measured at a gauging station (U. S. Army Corps of Engineers, 1991).

Groundwater pollution: Impairment of groundwater quality can result from land disposal of waste materials that are then dissolved by percolating water and leach downward through the unsaturated zone to the groundwater, or from improperly constructed or operated wells (After Flick, 1980, p. 83).

Groundwater recharge: Replenishment of groundwater naturally by precipitation or runoff or artificially by spreading or injection.

Groundwater reservoir: (1) An aquifer. (2) A term used to refer to all the rocks in the zone of saturation, including those containing permanent or temporary bodies of perched groundwater (AGE, 1980, p. 278).

Groundwater runoff: That portion of the runoff that has infiltrated the groundwater system and has later been discharged into a stream channel or other surface-water body as a spring or lake. Groundwater runoff is the principal source of base or dry weather flow of streams unregulated by surface storage, and such flow is sometimes called *groundwater flow* (Rogers et al., 1981, p. 174).

Gyben–Herzberg principle: An equation that relates the depth of a saltwater interface in a coastal aquifer to the height of the freshwater table above sea level (Fetter, 1994).

Half life: The time required for a pollutant to lose one-half of its original concentration (U.S. Environmental Protection Agency, 2001).

Hantush–Jacob formula: An equation to describe the change in hydraulic head with time during pumping of a leaky confined aquifer (Fetter, 1994).

Hardpan: A shallow layer of earth material that is relatively hard and impermeable, usually through the deposition of minerals (U. S. Army Corps of Engineers, 1991). *Cliche.*

Head, elevation: The elevation of a given point in a column of liquid above a datum (U. S. Army Corps of Engineers, 1991)

Heterogeneity: A characteristic of a medium in which material properties vary from point to point (U.S. Geological Survey, 1989). A lack of uniformity in porous media properties and conditions (Cohen and Mercer, 1993).

Heterogeneity: A characteristic of a medium in which material properties vary from point to point (Groundwater Subcommittee, 1989). Pertaining to a substance having different characteristics in different locations.

Hollow-stem auger: A particular kind of a drilling device whereby a hole can be rapidly advanced into sediment. Sampling and installation of the equipment can take place through the hollow center of the auger (Fetter, 1994).

Homogeneous: Of uniform composition throughout (U. S. Army Corps of Engineers, 1991).

Horizontal well: A tubular well pushed approximately horizontally into a water-bearing stratum or under the bed of a lake or stream. Also called *push well* (Rogers et al., 1981, p. 181).

Hot spring: Thermal spring, the water of which has a temperature higher than that of the human body, that is, higher than 98°F (36°C).

Hvorslev method: A procedure for performing a slug test in a piezometer that partially penetrates a water-table aquifer (Fetter, 1994).

Hydraulic conductivity: Property of an aquite rock that indicates its capability to transmit water, expressed as the volume of water at the existing kinematic viscosity that will move in unit time under a unit hydraulic gradient through a unit area of aquifer measured at right angles to the direction of flow (Lohman et al., 1988). *Permeability coefficient.*

Hydraulic diffusivity: Property of an aquifer or confining bed that is defined as the ratio of the transmissivity to the storativity (Fetter, 1994).

Hydraulic gradient: The change in head per unit distance in a given direction, typically in the principal flow direction (Cohen and Mercer, 1993). In a stream, the slope of the hydraulic grade line. The change in total head with a change in distance in a given direction. The direction is that which yields a maximum rate of decrease in head (Fetter, 1994).

Hydraulic head: The height above a datum plane (such as sea level) of the column of water that can be supported by the hydraulic pressure at a given point in a groundwater system (U. S. Department of the Interior, 1986).

Hydrochemical facies: Bodies of water with separate but distinct chemical compositions contained in an aquifer (Fetter, 1994).

Hydrogen-ion concentration: The negative log of the hydrogen-ion activity in solution, a measure of the acidity or basicity of a solution, commonly designated as pH (AGI, 1980, p. 470).

Hydrogeology: The study of the interrelationships of geologic materials and processes with water, especially groundwater (Fetter, 1994). Emphasis is on the geologic aspects of groundwater flow and occurrence.

Hydrograph: Graph showing stage, flow, velocity, or other property of water with respect to time (Langbein and Iseri, 1960, p. 12).

Hydrologic cycle: A convenient term to denote the circulation of water from the sea, through the atmosphere, to the land, and thence, with many delays, back to the sea by overland and subterranean routes and in part by way of the atmosphere; also the many short circuits of the water that are returned to the atmosphere without reaching the sea (Langbein and Iseri, 1960). The circulation of water from the sea, through the atmosphere, to the land, and thence, with many delays, back to the sea by overland and subterranean routes and in part by way of the atmosphere; also the many short circuits of the water that is returned to the atmosphere without reaching the sea.

Hydrologic data: Records of observations and measurements of physical facts, occurrences, and conditions related to precipitation, streamflow, groundwater, quality of water, and water use (After Rogers et al., 1981, p. 90).

Hydrologic equation: An expression of the law of mass conservation for purposes of water budgets. It may be as inflow equals outflow plus or minus changes in storage (Fetter, 1980).

Hydrologic map: Specialized map of which groundwater resources,

Hydrologic model: A scaled reproduction (mechanical device) or representation (electrical system) of hydrologic data (After Rogers et al., 1981, p. 235).

Hydrologic properties: Those properties of a rock that govern the entrance of water and the capacity to hold, transmit, and deliver water, for example, porosity, effective porosity, specific retention, permeability, and direction of maximum and minimum permeability (AGI, 1980, p. 302).

Hydrologic regime: The characteristic behavior and the total quantity of water involved in a drainage basin, determined by measuring such quantities as rainfall, surface and subsurface storage and flow, and evapotranspiration (AGI, 1980, p. 527).

Hydrologic system: A complex of related parts (physical, conceptual, or both) forming an orderly working body of hydrologic units, and their man-related aspects, such as the use, treatment, reuse, and disposal of water; the costs and benefits thereof; and the interaction of hydrologic factors with those of sociology, economics, and ecology (AGI, 1980, p. 302).

Hydrology: The science that relates to the water of the earth. The science treating the waters of the earth, their occurrence, distribution, and movement (Langbein and Iseri, 1960). The science encompassing the behavior of water as it occurs in the atmosphere, on the surface of the ground, and underground (American Society of Civil Engineers, 1949, p. 1). The study of the occurrence, distribution, and chemistry of all waters of the earth (Fetter, 1994).

Hypotheic zone: That portion of the saturated zone in surface water and groundwater mix (Weight, 2008).

Igneous rock: A rock that solidified from molten or partly molten material; one of the three main classes into which all rocks are divided (U.S. Geological Survey, 1993).

Image well: An imaginary well that can be used to simulate the effect of a hydro-logic barrier, such as a recharge boundary or a barrier boundary, on the hydraulics of a pumping or recharge well (Fetter, 1994).

Impervious bed: Bed or stratum through which water cannot move (Rogers et al., 1981, p. 192).

Impervious boundary: An aquifer boundary, such as is formed by a tight fault or impermeable wall of a buried stream valley that cuts off or prevents ground-water flow (After Ferris, Knowles, Brown, and Stallman, 1962, p. 145).

Induced infiltration: When a surface-water body is in hydraulic connection with an aquifer, and the cone of depression of a pumping well in that the aquifer reaches the water body, the source of some of the pumped water will be streamflow that is induced into the groundwater body under the influence of gradients set up by the well. When steady-state conditions are reached, the source of all the pumped groundwater is streamflow (Freeze and Cherry, 1979).

Industrial wastes: The wastes from industrial processes, as distinct from domestic or sanitary wastes (Rogers et al., 1981, p. 196).

Infiltration capacity: The maximum rate at which soil or aquifer, when in a given condition, can absorb infiltrating water.

Influent stream: See *losing stream.*

Inlet well: (1) Well or opening at the surface of the ground constructed to receive surface water, which is then conducted to a sewer. (2) A chamber that serves as a suction well for pumps in a wastewater pumping station (Rogers et al., 1981, p. 1990).

Inorganic: Being composed of material other than plant or animal materials (U. S. Army Corps of Engineers, 1991).

Instream water: Water that is used but not withdrawn from a groundwater or sur-face water source such as navigation and recreation (Vandas et al., 2002).

Interflow: The lateral movement of water in the unsaturated zone during and imme-diately after rainfall (Fetter, 1994).

Intermittent spring: A spring that discharges only periodically. A geyser is a spe-cial type of intermittent spring (Meinzer, 1923, p. 54).

Intrinsic permeability: A measure of the relative ease with which a porous medium can transmit a liquid under a potential gradient. Intrinsic permeability is a property of the medium alone that is dependent on the shape and size of the openings through which the liquid moves (Cohen and Mercer, 1993). Pertaining to the relative ease with which a porous medium can transmit a liquid under a hydraulic or potential gradient. It is a property of the porous medium and is independent of the nature of the liquid or the potential field (Fetter, 1994).

Intrusive rock: A rock that was emplaced as magma in preexisting rock.

Ion: An atom that has lost or gained one or more electrons and therefore has acquired an electrical charge.

Ion exchange: Reversible chemical replacement of an ion bonded at the liquid–solid interface by an ion in solution. A process by which an ion in a mineral lattice is replaced by another ion that was present in an aqueous solution (Fetter, 1944).

Production: It includes water required to satisfy surface evaporation and other economically unavoidable loss (Rogers et al., 1981, p. 206).

Isotropy: The ability of a fluid to move in a given direction at a given point within the aquifer (Weight, 2008).

Jacob straight-line method: A graphical method using semilogarithmic paper and the Theis equation for evaluating the results of an aquifer test (Fetter, 1994).

Junior water rights holder: One who holds rights that are temporarily more recent than senior rights holders. All water rights are defined in relation to other users, and a water rights holder only acquires the right to use a specific quantity of water under specified conditions. Thus, when limited water is available, junior rights are not met until all senior rights have been satisfied (U. S. Army Corps of Engineers, 1991).

Kaolinite: A common clay mineral of the kaolin group (After AGE, 1980, p. 337).

Karst: A terrain, generally underlain by limestone, in which the topography is formed by the dissolving rock and subsequent collapse, and which is commonly characterized by closed depression subterranean drainage and caves (After Monroe, 1970, p. K11). The type of geologic terrain underlain by carbonate rock where significant solution of the rock has occurred due to flowing groundwater (Fetter, 1994).

Karst hydrology: (1) The drainage phenomena of karstified limestone, dolomite, and other slowly soluble rocks (Monroe, 1970, p. K11). (2) The study of water occurring in karst.

Kriging: A statistical procedure that geologists use to characterize the subsurface. Kriging maximizes the information obtained from a given number of samples (National Academy, 1994).

Laminar flow: Flow in which the head loss is proportional to the first power of the velocity. That type of flow in which the fluid particles follow paths that are smooth, straight, and parallel to the channel walls. In laminar flow the viscosity of the fluid damps out turbulent motion (Groundwater Subcommittee, 1989). The type of flow in which the fluid particles follow paths that are smooth, straight, and parallel to the channel walls. Compare with *turbulent flow* (Fetter, 1994).

Landfill: A facility for the disposal of solid wastes or sludges by placing on land, compacting, and covering as appropriate with a thin layer of soil (Rogers et al., 1981, p. 213).

Landsat: An unmanned, earth-orbiting, National Aeronautics and Space Administration satellite that transmits multispectral images in the 0.4 to 1.1 mm region to Earth receiving stations. It was formerly called Earth Resource Technology Satellite, or ERTS (AGE, 1980, p. 349).

Landslide: A popular term used to describe all types of slope movement (Holzer, 2009).

Leachate: Water that contains a high amount of dissolved solids as a result of liquid seeping from a landfill (Fetter, 1994).

Leachate collection system: A system installed in conjunction with a liner to capture the leachate that may be generated from a landfill so that it may be taken away and treated (Fetter, 1994).

Leaching: (1) The removal of soluble constituents from soil, landfills, mine wastes, sludge deposits, or other material by percolating water. (2) The disposal of excess liquid through porous soil or rock strata (Rogers et al., 1981, p. 214).

Leakage: (1) The uncontrolled loss of water from artificial structures as a result of hydrostatic pressure. (2) The uncontrolled loss of water from one aquifer to another. The leakage may be natural, as through a semipervious confining layer, or man-made, as through an uncased well (Rogers et al., 1981, p. 215).

Leakance: The rate of flow across a unit (horizontal) area of a semipervious layer into (or out of) an aquifer under one unit of head difference across this layer (U. S. Department of the Interior, 1989). The ratio K'/b' in which K' and b' are the vertical hydraulic conductivity and the thickness, respectively, of the confining bed (ASTM, 1994). *Vertical hydraulic conductivity.*

Leaky aquifer: Confined aquifer whose confining beds will conduct significant quantities of water into or out of the aquifer (AGE, 1980, p. 355).

Leaky confining bed: A confining bed that transmits water at sufficient rates to furnish some recharge to a well pumping from an underlying aquifer (Fetter, 1994). A confining bed that under natural conditions transmits water vertically between two aquifers because of hydraulic head differences between aquifers.

Limestone: A consolidated sedimentary rock composed largely of the carbonate minerals calcite or, less frequently, aragonite. Limestones are important as aquifers, as reservoir rocks for hydrocarbons, as building stone and aggregate, and with clay, for making cement (After Allaby, 1977, p. 291).

Lithologic log: A record of the sequence of lithologic characteristics of the rocks penetrated in drilling a well, compiled from examination of well cuttings and cores. The information is referred to as depth of origin and is plotted on a strip log form (AGI, 1980, p. 553).

LNAPL: An acronym for less-dense-than-water nonaqueous phase liquid. LNAPLs do not mix well with water and are less dense than water. Gasoline and fuel oil are common LNAPLs.

Loam: Soil containing 7 to 27 percent clay, 28 to 50 percent silt, and less than 52 percent sand (American Society of Agricultural Engineers, 1967, p. 21).

Losing stream: Stream or reach of stream that contributes water to an aquifer. The water surface of such a stream stands at a higher level than the water table or potentiometric surface of the groundwater body to which it contributes water (Rogers et al., 1981, p. 198).

Magnetometer: An equation that can be used to locate items that disrupt the earth's localized magnetic field; can be used for finding buried steel (Fetter, 1994).

Maximum contaminant level (MCL): The highest concentration of a solute permissible in a public water supply as specified in the National Interim Primary Drinking Water standards for the United States (Fetter, 1994).

Measuring point: An arbitrary permanent reference point from which the distance to the water surface in a well is measured to obtain the water level (Novak, 1985).

Metamorphic rock: Any rock derived from preexisting rocks in response to marked changes in temperature, pressure, shearing stress, and chemical environment at depth in the earth's crust; one of the three main classes into which all rocks are divided.

Micrograms per liter: A measure of the amount of dissolved solids in a solution in terms of micrograms of solute per liter of solution (Fetter, 1994).

Mineral water: Water that contains a large quantity of mineral salts (Rogers et al., 1981, p. 234).

Model: A conceptual, mathematical, or physical system obeying certain specified conditions, whose behavior is used to understand the physical system to which it is analogous in some way (Groundwater Subcommittee, 1989).

Model calibration: The process by which the independent variables of a digital computer model are varied in order to calibrate a dependent variable such as hydraulic head against observed water-table elevations (Fetter, 1994).

Modeling: The simulation of some physical or abstract phenomenon or system with another system believed to obey the same physical laws or abstract rules of logic, in order to predict the behavior of the former (main system) by experimenting with the latter (analogous system; Rogers et al., 1981, p. 235).

Model verification: The process by which a digital model that has been calibrated against a steady-state condition is tested to see if it can generate a transient response, such as the decline in the water table with pumping, that matches the known history of the aquifer (Fetter, 1994).

Monitoring: A type of sampling program designed to determine time trend changes in water level, water quality, and streamflow (After U. S. Army Corps of Engineers, 1991)

NAPL: Nonaqueous phase liquids. A liquid solution that does not mix easily with water. Many common groundwater contaminants, including chlorinated solvents and many petroleum products, enter the subsurface in the nonaqueous phase solutions (National Academy, 1993).

Natural water table: A water table in its natural condition and position, not disturbed by artificial additions or extractions of water (Rogers et al., 1981, p. 243).

Natural well: An abrupt depression in the land surface, not made by human activity, which extends into the saturation zone but from which water does not flow to the surface except by artificial processes. It is distinguished from such features as ponds, swamps, lakes, and other bodies of impounded surface water, which also extend into the saturation zone, by having a smaller water surface, being deeper in proportion to its water surface area, and having steeper sides. Although it is not a well in a strict sense, the term is well established as applying to such a feature (Rogers et al., 1981, p. 243).

Neutron log: A curve that indicates variations in the intensity of radiation (neutrons or gamma rays) with depth produced when the rocks penetrated by a borehole are bombarded by neutrons from a sonde. It indicates the presence of fluids (but does not distinguish between oil and water) in the rocks, and is used with the gamma-ray log to differentiate porous from nonporous formations (AGE, 1980, p. 422).

Nitrate: Mineral compound characterized by a fundamental anionic structure of NO_3. Soda niter, $NaNO_3$, and niter, KNO_3, are nitrates (AGI, 1980, p. 424).

Nonpoint pollution source: Natural or man-induced alteration of the chemical, physical, biological, or radiological integrity of water, originating from any source other than a point source (After Rogers et al., 1981, p. 247).

Nonpotable: Water that is unsafe or unpalatable to drink because it contains pollutants, contaminants, minerals, or infective agents (U.S. Environmental Protection Agency, 2001).

Nonuniform flow: A flow in which the slope, cross-sectional area, and velocity change from section to section in the channel (Rogers et al., 1981, p. 248).

Numerical model: A model whose solution must be approximated by varying the values of controlling parameters and using computers to solve approximate forms of model's governing equations (National Academy, 1994). The analytical model uses classical mathematical tools, such as differential equations.

Observation well: A nonpumping well used to observe the elevation of the water table or potentiometric surface (Fetter, 1994).

Outcrop: That part of the formation that appears at the surface of the ground (U. S. Army Corps of Engineers, 1991).

Outwash: Stratified detritus (chiefly sand and gravel) carried away from a glacier by meltwater streams and deposited in front of or beyond the end moraine or the margin of an active glacier. The material that is deposited near the ice is coarser than that deposited farther downstream (After AGE, 1980, p. 446).

Packer test: An aquifer test performed in an open borehole; the segment of the aquifer to be tested is sealed off by inflating seals in the borehole, called packer, both above and below the segment (After Fetter, 1960).

Partially penetrating well: Well that does not penetrate the entire thickness of the aquifer (Fetter, 1994).

Particle size: A linear dimension, usually designated as *diameter*, used to characterize the size of a particle. The dimension may be determined by any of several different techniques, including seimentation, siefing, midrometric measurement, or direct measurement. *Grain size.*

Pathogen: Pathogenic or disease-producing organism (Rogers et al., 1981, p. 262).

Pathogenic bacteria: Bacteria that cause disease in the host organism by their parasitic growth (Rogers et al., 1981, p. 262).

PCB: An acronym of polychlorinated biphenyl compounds. PCBs are extremely stable, nonflammable, dense, and viscous liquids that are formed by substituting chlorine atoms for hydrogen atoms on a biphenyl molecule. PCBs were manufactured by Monsanto Chemical Company for use as dielectric fluids for electrical transformers (Cohen and Mercer, 1993).

Perched groundwater: Groundwater separated from an underlying body of groundwater by an unsaturated zone. Its water table is a perched water table. Perched groundwater is held up by a perching bed whose permeability is so low that water percolating downward through is not able to bring water in the underlying unsaturated zone above atmospheric pressure (Groundwater Subcommittee, 1989).

Perched spring: Spring whose source of water is a body of perched groundwater (AGE, 1980, p. 465).

Perched water table: The water table of a body of perched groundwater (AGE, 1980, p. 465).

Perennial spring: A spring that flows continuously, as opposed to an intermittent spring (AGE, 1980, 465).

Permafrost: Permanently frozen ground (Holzer, 2009).

Permeability: The property of soil or rock that allows passage of water through it when subjected to a difference in head. This depends not only on the volume of the openings and pores but also on how these openings are connected to each other (U. S. Army Corps of Engineers, 1991). The property or capacity of a porous rock, sediment, or soil for transmitting a fluid; it is a measure of the relative ease of fluid flow under unequal pressure. The customary unit of measurement is the millidarcy (AGE, 1980, p. 468).

Permeability coefficient: The rate of flow of water in gallons per day through a cross-section of one square foot under a unit hydraulic gradient, at the prevailing temperature (field permeability coefficient) or adjusted for a temperature of 60°F (AGE, 1980, p. 468).

Permit system: A general term referring to a system of acquiring water rights under state law whereby the state must issue a permit for a new use of water; although permit systems were, at one time, generally associated with eastern states using the riparian doctrine, they are now found in the South and West.

Pesticide: Any substance, organic or inorganic, used to kill plant or animal pests; major categories of pesticides include herbicides and insecticides.

pH: See *hydrogen-ion concentration*.

Phosphate: Salt or ester of phosphoric acid.

Phreatic: Of or pertaining to groundwater.

Phreatic decline: See *water-table decline*.

Phreatophyte: Plant that obtains its water supply from the water table or from the capillary fringe and is characterized by a deep root system (AGE, 1980, p. 473).

Physical analysis: The examination of water and wastewater to determine physical characteristics such as temperature, turbidity, color, odors, and taste (Rogers et al., 1981, p. 270).

Piezometer: Instrument for measuring pressure head in a conduit, tank, or soil. Usually consists of a small pipe or tube tapped into the side of the container, with its inside end flush with, and normal to, the water face of the container, and connected with a manometer pressure gauge, mercury or water column, or other device for indicating pressure head (Rogers et al., 1981, p. 271).

Piezometer nest: A set of two or more piezometers set close to each other but screened to different depths (Fetter, 1994).

Piezometric head: The elevation plus the pressure head. Above a datum, the total head at any cross-section minus the velocity head at that cross-section. It is equivalent to the elevation of the water surface in open channel flow; it is the elevation of the hydraulic grade line at any point (Rogers et al., 1981, p. 271).

Pitless adaptor: A fitting from the riser pipe discharge line into a welded port in the casing allowing water to be discharged in a water line to the house. They are placed below the frost line.

Playa: Dry, vegetation-free, flat area at the lowest part of an undrained desert basin, underlain by stratified clay, silt, or sand, and commonly by soluble salts. The term is also applied to the basin containing an expanse of playa, which may be marked by ephemeral lakes (AGE, 1981, p. 483). Broadly used to describe areas occupied by temporary shallow lakes that are the focus of an area of internal drainage. They usually occur in arid and semi-arid areas. Playa lakes are not part of an integrated surface drainage system. They are found in New Mexico, Utah, Nevada, and Texas (Stone and Stone, 1994).

Plume: A zone of dissolved contaminants. A plume usually will originate from the DNAPL zone and extends downgradient for some distance, depending on site hydrogeologic and chemical conditions (Cohen and Mercer, 1993). A plume usually will originate from the contaminant source areas and extend downgradient for some distance, depending on site hydrogeologic and chemical conditions (National Academy, 1994).

Point source of pollution: Pollution originating from any discrete source, such as the outflow from a pipe, ditch, tunnel, well, concentrated animal-feeding operation, or floating craft.

Pollution: The condition caused by the presence of substances of such character and in such quantities that the quality of the environment is impaired. Presence of substance in the water that is or could become injurious to the public health, safety, or welfare; or that is or could become injurious to domestic, commercial, industrial, agriculture, or other uses being made of the water.

Pore: The volume between mineral grains that is occupied by fluid in a porous medium. A small space between the grains of sand or soil. Groundwater is stored in pores.

Pore pressure: The stress transmitted by the fluid that fills the voids between particles of a soil or rock mass, for example, that part of the total normal stress in a saturated soil caused by the presence of interstitial water (AGE, 1980, p. 422.)

Pore water: *Interstitial water.*

Porosity: Property of containing interstices or voids; may be expressed quantitatively as the ratio of the volume of interstices to the total volume of material as either a decimal fraction or percentage (Meinzer, 1923, p. 19).

Porous: Having numerous interstices, whether connected or isolated; usually refers to openings of smaller size than those of a cellular rock (AGE, 1980, p. 492).

Potable water: Water that does not contain objectionable pollution, contamination, minerals, or infective agents and is considered satisfactory for domestic consumption (Rogers et al., 1981, p. 279).

Potential evaporation: *Evapotranspiration potential.*

Potentiometric surface: The surface that represents the level to which water will rise in tightly cased wells. If the head varies significantly with depth in the aquifer, then there may be more than one potentiometric surface. The water table is a particular potentiometric surface (Fetter, 1994).

Precipitation: As used in hydrology, the discharge of water, in liquid or solid state, out of the atmosphere, generally upon a land or water surface (Langbein and Iseri, 1960); water that falls to the surface from the atmosphere as rain, snow, hail. or sleet. It is measured as a liquid-water equivalent regardless of the form in which it falls (AGE, 1980, p. 496).

Precipitation gauge: *Rain gauge.*

Pressure gauge: A device for registering the pressure of solids, liquids, or gases. It may be graduated to register pressure in any units desired (Rogers et al., 1981, p. 282). *Manometer, piezometer.*

Pressure head: The head represented by the expression of pressure over weight. The head is usually expressed as height of liquid in a column corresponding to the weight of the liquid per unit area, for example, feet head of water corresponding to pounds per square inch (Rogers et al., 1981, p. 282).

Price current meter: A current meter with a series of conical cups fastened to a flat framework through which a pin extends. The pin sits in the framework of the meter, and the cups are rotated around it in a horizontal plane by the flowing water, registering the number of revolutions by acoustical or electrical devices, from which the velocity of the water may be computed (U. S. Army Corps of Engineers, 1991).

Primary porosity: The porosity that represents the original pore openings when a rock or sediment is formed (Fetter, 1994).

Prior-appropriation doctrine: A concept in water law stating that the right to use water is separate from other property rights and that the first person to withdraw and use the water holds the senior right. The doctrine has been applied to both surface and groundwater (After Fetter, 1994).

Pulsed pumping: An enhancement of the pump and treat system in which extraction wells are periodically not pumped.

Pump: A mechanical device for causing flow, for raising or lifting water or other fluid, or for applying pressure to a fluid (Rogers et al., 1981, p. 288).

Pump and treat system: Most commonly used type of system for cleaning up contaminated groundwater. Pump and treat systems consist of a series of wells used to pump contaminated water to the surface and a surface treatment facility used to clean the extracted groundwater (National Academy, 1993).

Pumpage: The total quantity of liquid pumped in a given interval, usually a day, a month, or a year (Rogers et al., 1981, p. 288).

Pumped well: Well that discharges water at the surface through the operation of a pump or other lifting device (Rogers et al., 1981, p. 290).

Pumping test: A test made by pumping a well for a period of time and observing the change in hydraulic head in the aquifer. A pumping test may be used to determine the capacity of the well and the hydraulic properties of the aquifer (Fetter, 1994). See *aquifer test.*

Qanat: A term used in the Middle East, primarily in Iran, for an ancient, gently inclined underground channel or conduit dug so as to conduct groundwater by gravity from alluvial gravel at the foot of hills to an arid lowland; a type of horizontal well (After AGE, 1980, p. 512).

Quality assurance: Programs and sets of procedures including but not limited to quality control that are used to ensure product quality or data reliability.

Quality control: Procedures used to regulate measurements and produce data that meet the needs of the user.

Radial flow: The flow of water in an aquifer toward a vertically oriented well (Fetter, 1994).

Radial well: A special adaption of infiltration galleries when the well screen extends horizontally from a large vertical caisson constructed adjacent to a stream, river, or lake.

Radioactivity: The spontaneous decay or disintegration of an unstable atomic nucleus, accompanied by the emission of radiation.

Radionuclide: A species of atom that emits alpha, beta, or gamma rays for a measurable length of time. Radionuclide is distinguished by atomic weight and atomic number.

Radius of influence of a well: Distance from the center of the well to the closest point at which the potentiometric surface is not lowered when pumping has produced the maximum steady rate of flow.

Rain: Particles of liquid water that have become too large to be held by the atmosphere. Their diameter generally is greater than 0.02 inches (0.5 mm) and they usually fall to the earth at velocities greater than 10 feet per second (3.05 m/s) in still air (Rogers et al., 1981, p. 293).

Rating curve: A graph of the discharge of a river at a particular point as a function of the elevation of the water surface (Fetter, 1994).

RCRA: Acronym for Resources Conservation and Recovery Act, which regulates monitoring, investigation, and corrective action activities at all hazardous treatment, storage, and disposal facilities (Cohen and Mercer, 1993).

Recharge: (1) Addition of water to the zone of saturation from precipitin, infiltration from surface streams, and other sources. (2) To inject water under ground to replenish groundwater and/or prevent land collapse or saltwater intrusion (Rogers et al., 1981, p. 299).

Recharge area: An area in which water reaches the zone of saturation by surface infiltration. An area in which there are downward components of hydraulic head in the aquifer. Infiltration moves downward into the deeper parts of an aquifer in a recharge area (Fetter, 1994).

Recharge well: Well constructed to conduct surface water or other surplus water into an aquifer to increase the groundwater supply. Sometimes called *diffusion well* (Rogers et al., 1981, p. 300). A well specifically designed so that water can be pumped into an aquifer in order to recharge the groundwater reservoir (Fetter, 1994).

Redox potential: Describes the distribution of oxidized and reduced species in a solution at equilibrium; important for predicting the likelihood that metals will precipitate from groundwater upon pumping, for estimating the capacity of microorganisms to degrade contaminants, and for predicting other subsurface reactions.

Recovery: The rate or amount of water-level rise in a well after the pump has been shut off. It is the inverse of *drawdown* (Fetter, 1994).

Relief well: A well used to relieve excess hydrostatic pressure, in order to reduce waterlogging of soil or to prevent blowouts on the land side of a levee or dam at times of high water (AGE, 1980, p. 530).

Remote sensing: The collection of information about an object by a recording device that is not in physical contact with it. The term is usually restricted to mean methods that record reflected or radiated electromagnetic energy rather than methods that involve significant penetration into the Earth. The technique employs such devices as the camera, infrared detectors, microwave frequency receivers, and radar systems (AGE, 1980, p. 530).

Reparian zone: Commonly referred to as the interface between the terrestrial and aquatic interface (Weight, 2008).

Residual drawdown: The difference between the projected prepumping water-level trend and the water level in a well or piezometer after pumping or injection has stopped (ASTM, 1994).

Resistivity survey: Any electrical exploration method in which current is introduced into the ground by two contact electrodes and potential differences are measured between two or more other electrodes (AGE, 1980, p. 532).

RFI: Acronym for RCRA Facility Investigation.

River: (1) A general term for a natural freshwater surface stream of considerable volume and a permanent or seasonal flow, moving in a definite channel toward a sea, lake, or another river; any large stream, or one larger than a brook or a creek, such as the trunk stream and the larger branches of a drainage system. (2) A term applied in New England to a small watercourse that elsewhere in the United States is known as a creek (AGE, 1980, p. 541). *Stream.*

Safe Drinking Water Act: A law passed in 1974 that required the setting of standards to protect the public from exposure to contaminants in drinking water.

Safe yield: The rate at which water can be withdrawn from an aquifer for human use without depleting the supply to such an extent that withdrawal at this rate is no longer economically feasible (Meinzer, 1923, p. 55). Use of the term is discouraged because the feasible rate of withdrawal depends on the location of wells in relation to aquifer boundaries and rarely can be estimated in advance of development (AGE, 1980, p. 550).

Saline aquifer: Aquifer containing salty water (AGE, 1980, p. 551).

Saline-water system: A system of water mains, separate from the regular distribution system, that conveys salt water for fighting fires (Rogers et al., 1981, p. 319).

Salinity: (1) The relative concentration of dissolved salts, usually sodium chloride, in a given water. It is often expressed in terms of mg/L chlorine. (2) A measure of the concentration of dissolved mineral substances in water (Rogers et al., 1981, p. 318).

Sample log: See *lithologic log.*

Sampling: The systematic gathering of specimens for appraisal. Such appraisal may consist of the chemical, physical, or biological content of the specimen.

Sand: (1) A loose material consisting of small but easily distinguishable grains, most commonly of quartz resulting from disintegration of rocks. (2) Sediment particles having diameters between 0.062 and 2.0 mm. (3) A size classification of sediment transported by water, air, or ice (Rogers et al., 1981, p. 320).

Sandstone: Consolidated sedimentary rock composed mainly of cemented sand and recognized by a gritty feeling; depending on mineral content, it may be referred to as quartzose sandstone (mainly quartz), arkosic sandstone (>20% feldspar), or graywacke (poorly sorted sandstone).

Sanitary landfill: A solid-waste burial site operated to minimize environmental hazards. The wastes are continuously compacted and are covered daily with a layer of soil to minimize blowing, odors, fire hazards, and fly and rat problems; provisions are also made to minimize leaching of dissolved solids to groundwater (Rogers et al., 1981, p. 321).

Saturated thickness: The thickness of saturated permeable material in an unconfined aquifer from the water table to the lowermost confining layer (Weight, 2008).

Saturated zone: The zone in which the voids in the rock or soil are filled with water at a pressure greater than atmospheric. The water table is the top of the saturated zone in an unconfined aquifer (Fetter, 1994). See *zone of saturation*.

Screen: A device with openings, generally having a relatively uniform size, that permit liquid to pass but retain larger particles. The screen may consist of bars, coarse to fine wire, rods, gratings, membranes, and so forth, depending upon particle size to be retained (U. S. Army Corps of Engineers, 1991).

Screened well: Well into which water enters through one or more screens (Rogers et al., 1981, p. 323).

Sea level: In general, the surface of the sea used as a reference for elevation. A curtailed form of mean sea level (Rogers et al., 1981, p. 325).

Secondary porosity: The porosity that has been caused by fractures or weathering in a rock or sediment after it has been formed (Fetter, 1994).

Sedimentary rock: A rock resulting from the consolidation of loose sediment that has accumulated in layers; a general term for any sedimentary material, unconsolidated or consolidated (AGI, 1980).

Seep: (1) A poorly defined area where water oozes from the earth in small quantities. (2) To appear or disappear, as water or other liquid, from a poorly defined area of the Earth's surface. (3) According to some authorities, the type of movement of water in unsaturated material (Rogers et al., 1981, p. 329).

Seep water: Water that has passed through or under a levee or dam (Rogers et al., 1981, p. 330).

Seepage loss: The loss of water by capillary action and slow percolation (Rogers et al., 1981, p. 329).

Seepage spring: Spring in which the water percolates from numerous small openings in permeable material. Also called *filtration spring* (Rogers et al., 1981, p. 329).

Seismic: Pertaining to an earthquake or Earth vibration, including one that is artificially induced (AGE, 1980, p. 568).

Seismic refraction: A method of determining subsurface geophysical properties by measuring the length of time it takes for artificially generated seismic waves to pass through the ground (Fetter, 1994).

Septic tank: An underground vessel for treating wastewater from a single dwelling or building by a combination of settling and anaerobic digestion. Effluent is usually disposed of by leaching. Settled solids are pumped out periodically and hauled to a treatment facility for disposal (Rogers et al., 1981, p. 332).

Shale: A thinly bedded sedimentary rock consisting mostly of clay-sized grains (Holzer, 2009).

Sink: General term for a closed depression. It may be shaped like a basin, funnel, or cylinder (Monroe, 1970, p. 16).

Site investigation: The collection of basic facts and the testing of surface and subsurface materials (including their physical properties, distribution, and geologic structure) at a site, for the purpose of preparing suitable designs for an engineering structure or other uses (AGE, 1980, p. 585).

Slug test: An aquifer test made either by instantaneously adding a volume of water into a well or by instantaneously withdrawing a volume of water from the well. A synonym for this test, when a slug of water is removed from the well, is a bail-down test (Fetter, 1994).

Soil: Unconsolidated mineral and organic surface material that has been sufficiently modified and acted upon by physical, chemical, or biological agents so that it will support plant growth.

Soil type: A phase or subdivision of a soil series based primarily on texture of the surface soil to a depth at least equal to plow depth (about 15 cm). In Europe, the term is roughly equivalent to the term *great soil group* (AGE, 1980, p. 593).

Soil water: Water diffused in the soil; the upper part of the zone of aeration from which water is discharged by the transpiration of plants or by direct evaporation (Langbein and Iseri, 1960, p. 17). *Soil moisture.*

Soil-water table: The upper surface of a body of soil water (Rogers et al., 1981, p. 354).

Specific capacity: The rate at which water may be drawn from a formation through a well to cause a drawdown. The usual units of measurement are gallons per minute per foot of drawdown (U. S. Army Corps of Engineers, 1991). An expression of the productivity of a well, obtained by dividing the rate of discharge of water from the well by the drawdown of the water level in the well. Specific capacity should be described on the basis of the number of hours of pumping prior to the time the drawdown measurement is made. It will generally decrease with time as the drawdown increases (Fetter, 1994).

Specific conductance: A measure of the ability of a sample of water to conduct electricity. Measurement in micromohs. Used to estimate total dissolved solid content in water (U.S. Geological Survey, 1993, p. 533).

Specific conductivity: With reference to the movement of water in soil, a factor expressing the volume of transported water per unit of time in a given area (AGE, 1980, p. 598).

Specific discharge: An apparent velocity calculated from Darcy's Law; represents the rate at which water would flow in an aquifer if the aquifer were an open conduit (Fetter, 1994).

Specific gravity: The ratio of a substance's density to the density of the same standard substance, usually water (Cohen and Mercer, 1993).

Specific weight: The weight of a substance per unit of volume. The units are newtons per cubic meter (Fetter, 1994).

Specific yield: The ratio of the volume of water that a given mass of saturated rock or soil will yield by gravity to the volume of that mass. The ratio is stated as a percentage (AGE, 1980, p. 598).

Split-spoon sample: A sample of unconsolidated material taken by driving a sampling device ahead of the drill bit in a boring (Fetter, 1994).

Spontaneous potential log: A borehole log made by measuring the natural electrical potential that develops between the formation and the borehole fluids (Fetter, 1994).

Spring: A place where groundwater flows naturally from a rock or the soil onto the land surface or into a body of surface water. Its occurrence depends on the nature and relationship of rocks, especially permeable and impermeable strata, on the position of the water table, and on the topography (AGE, 1980, p. 604).

Spring water: Water derived from a spring (Rogers et al., 1981, p. 361).

Step drawdown test: The test was developed to examine the performance of a well. In the test, the well is pumped at several successively higher pumping rates and the drawdown for each step is recorded.

Stiff pattern: A graphical means of presenting the chemical analysis of major cations and anions in a water sample (Fetter, 1994).

Storage coefficient: The volume of water an aquifer releases from or takes into storage per unit surface area of the aquifer per unit change in head. In a confined water body, the water derived from storage with decline in head comes from expansion of the water and compression of the aquifer; similarly, water added to storage with a rise in head is accommodated partly by compression by expansion of the aquifer. In an unconfined water body, the amount of water derived from or added to the aquifer by these processes generally is negligible compared with that involved in gravity drainage or filling of pores; hence, in an unconfined water body the storage coefficient is virtually equal to the specific yield (Lohman et al., 1972, p. 13). *Storativity.*

Stream gauge: A device for measuring the elevation of the water surface in a channel, ordinarily by the position of a float in a tube, the change in pressure at some datum, or a visual reading on stakes or boards along channel sides. The datum is commonly arbitrary, as is the case where zero is an estimate, possibly many years old, of the lowest stage probable (Russell, 1968, p. 83).

Submarine spring: Large offshore emergence of freshwater, usually associated with a coastal karst area but sometimes with lava tubes (AGE, 1980, p. 623).

Submersible pump: Centrifugal pump, usually electrical, capable of operating entirely submerged under water.

Subsurface water: Water in the lithosphere. It may be in liquid, solid, or gaseous state. It comprises unsatured zone water and groundwater (Rogers et al., 1981, p. 378).

Surficial aquifer: An aquifer near the earth's surface, in the most recent of geologic deposits.

Suspended load: The sediment load that is suspended sediment (Interagency Committee on Water Data, 1977). Sediment that is collected by standard sediment samplers.

Test hole: Circular hole made by drilling to explore for valuable minerals or to obtain geological or hydrological information (After AGE, 1980, p. 188).

Test well: A well constructed for test purposes as opposed to operational water-supply purposes.

Theis equation: An equation that describes the flow of groundwater in a fully confined aquifer under certain idealized conditions (Fetter, 1994).

Theissen method: A process used to determine the effective uniform depth of precipitation over a drainage basin with a nonuniform distribution of rain gauges (Fetter, 1994).

Trace element: A chemical element dissolved in water in minute quantities, always or almost always in concentrations less than 1 mg of trace element in 1 L of water (U.S. Geological Survey, 1993, p. 584).

Tracer: (1) A foreign substance mixed with or attached to a given substance for subsequent determination of the location or distribution of the substance. (2) An element or compound that has been made radioactive so that it can be easily followed (traced) in biological and industrial processes. Radiation emitted by the radioisotope pinpoints its location (Rogers et al., 1981, p. 398).

Tracer test: A method used to determine the flow of groundwater in the subsurface (National Academy, 1994).

Transmissivity: The rate at which groundwater of the prevailing kinematic viscosity is transmitted through a unit width of aquifer under a unit hydraulic gradient (Lohman et al., 1972, p. 13). It is a function of properties of the liquid, the porous media, and the thickness of the porous media (Fetter, 1994).

Transpiration: The quantity of water absorbed and transpired and used directly in the building of plant tissue, in a specified time. It does not include soil evaporation (U. S. Army Corps of Engineers, 1991). The process by which plants give off water vapor through their leaves (Fetter, 1994).

Tributary: Stream or other body of water, surface or underground, which contributes its water to another and larger stream or body of water (Rogers et al., 1981, p. 401).

Unconfined aquifer: An aquifer having a water table; an aquifer containing unconfined groundwater (AGE, 1980, p. 675). *Water-table aquifer.*

Underflow: The downstream flow of groundwater through permeable deposits that underlie a stream and that are more or less limited by rocks of lower permeability (Langbein and Iseri, 1960, p. 201).

Unsaturated flow: Flow of water in the unsaturated zone. Movement of water in a soil, the pores of which contain both air and water (Hutchinson, 1970, p. 50).

Unsaturated zone: That part of the lithosphere in which the functional interstices of permeable rock or earth are not (except temporarily) filled with water under hydrostatic pressure; that is, the water in the interstices is held by

Unsteady flow: Flow in which the velocity and the quantity of water flowing per unit time at every point along the conduit vary with respect to time and position (Rogers et al., 1981, p. 408).

Upconing: (1) The cone-shaped rise of salt water beneath freshwater in an aquifer as freshwater is produced from a well. (2) The cone-shaped rise of water underlying oil or gas in a reservoir as the oil or gas is withdrawn from a well (AGE, 1980, p. 133).

Vacuum extraction: The forced extraction of gas (with volatile contaminants) from the vadose zone, typically to prevent uncontrolled migration of contaminated soil gas and augment a site cleanup (Cohen and Mercer, 1993).

Vadose water: Subsurface water that partially occupies interstices in the zone of aeration. It comprises hygroscopic, pellicular, mobile, and fringe water (Rogers et al., 1981, p. 384).

Vadose zone: See *unsaturated zone.*

Venting, soil: Removes volatile organic compounds from the unsaturated zone by vacuum pumping, which causes large volumes of air to circulate through the spill. The presence of the circulating air promotes volatilization and possibly biodegration (Dominico and Schwartz, 1990).

Verification: The process of testing a model to an observed set of data using the model parameters derived during calibration. Model calibration and verification must be performed on separate sets of data.

Virus: The smallest (10 to 300 mu in diameter) life form capable of producing infection and diseases in man or other large species (Rogers et al., 1981, p. 414).

Volatile organic compounds (VOC): A group of lightweight synthetic organic compounds, many of which are aromatic; sometimes referred to as *purgeable organic compounds* because of their low solubility in water (U.S. Geological Survey, 1986, p. 552).

Volcanic spring: Spring with water derived at considerable depths and brought to the surface by volcanic forces (Rogers et al., 1981, p. 415).

Volcano: (1) Vent in the surface of the Earth through which magma and associated gases and ash erupt; also, the form or structure, usually conical, that is produced by the ejected material. (2) Any eruption of material, for example, mud, that resembles a magmatic volcano (AGE, 1981, p. 690).

Wadi: A valley, ravine, or watercourse that is dry except during the rainy season. Term usually used in reference to northern African and Arabian locales (Rogers et al., 1981, p. 416). *Arroyo.*

Water hole: (1) Natural hole or hollow containing water; a hole in the dry bed of an intermittent stream; a spring in a desert; also, a pool, pond, or small lake (Rogers et al., 1981, p. 423). (2) A natural or artificial pool, pond, or small lake that stores water.

Water level: (1) Water surface elevation or stage (U. S. Army Corps of Engineers, 1991). (2) The free surface of a body of water. (3) The elevation of the free surface of a body of water above or below any datum (Rogers et al., 1981, p. 423). (4) The surface of water standing in a well is usually indicative of the position of the water table or other potentiometric surface.

Water-level fluctuation: (1) Movement of the free-water surface of a body of water. (2) Change in the elevation of the free-water surface of a body of water above or below any datum (Rogers et al., 1981, p. 423).

Water-level recorder: Device for producing, graphically or otherwise, a record of the rise and fall of a water surface with respect to time (Rogers et al., 1981, p. 423).

Water resources: The supply of ground and surface water in a given area.

Water requirement: The quantity of water, regardless of its source, required by a crop in a given period of time for its normal growth under field conditions. It includes surface evaporation and other economically unavoidable states.

Water reuse: Application of appropriately treated wastewater to a constructive purpose (Rogers et al., 1981, p. 308).

Water right: The legal right to use a specific quantity of water, on a specific time schedule, at a specific place, and for a specific purpose (National Academy, 1992).

Water-supply system: (1) Collectively, all property involved in a water utility, including land, water source, collection systems, wells, dams and hydraulic structures, water lines and appurtenances, pumping systems, treatment works, and general properties. (2) In plumbing, the water distribution system in a building and its appurtenances (Rogers et al., 1981, p. 21).

Water table: The upper surface of the zone of saturation. The surface in an unconfined aquifer or confining bed at which the pore-water pressure is atmospheric. Its position can be identified by measuring the water level in shallow wells extending a few feet into the zone of saturation (Meinzer, 1923, p. 22; Langbein and Iseri, 1960, p. 21). It can be measured by installing shallow wells extending a few feet into the zone of saturation and then measuring the water level in those wells (Fetter, 1994).

Water-table aquifer: See *unconfined aquifer.*

Water-table decline: The downward movement of the water table. Also called *phreatic decline* (Rogers et al., 1981, p. 429).

Water-table fluctuation: The alternate upward and downward movement of the water table. Also called *phreatic fluctuation* (Rogers and others, 1981, p. 429).

Water-table gradient: The rate of change of altitude per unit of distance in the water table at a given place and in a given direction. If the direction is not mentioned, it is generally understood to be the direction along which the maximum rate of change occurs. Where the rate of change is uniform between two points, the gradient is equal to the ratio of the difference of altitude between the two points to the horizontal distance between them (Rogers et al., 1981, p. 429).

Water-table map: A specific type of potentiometric surface map for an unconfined aquifer; shows lines of equal elevation of the water table (Fetter, 1994).

Water treatment: (1) The filtration or conditioning of water to render it acceptable for a specific use. (2) A series of chemical and physical processes to remove dissolved and suspended solids from a raw water to produce potable water for distribution and use (Rogers et al., 1981, p. 430). *Purification.*

Water use: A system of classifying the purposes for which water is used, such as potable water supply, recreation and bathing, fish culture, industrial processes, waste assimilation, transportation, and power production (After Rogers et al., 1981, p. 430).

Water yield: The runoff from the drainage basin including groundwater outflow that appears in the stream plus groundwater outflow that bypasses the gauging station and leaves the basin underground; precipitation minus evapotranspiration (U. S. Army Corps of Engineers, 1991).

Well: (1) Artificial cylindrical excavation that derives fluid (oil, gas, or water) from the saturated zone (Rogers et al., 1981, p. 433). (2) An artificial excavation or shaft that collects water from the interstices of the soil or rock.

Well capacity: The maximum rate at which a well will yield water under a stipulated set of conditions, such as a given drawdown, pump and motor, or engine size (Rogers et al., 1981, p. 434).

Well casing: Material, usually pipe, used to line the borehole of a well (Rogers et al., 1981, p 434). A solid piece of pipe, typically steel or PVC plastic, used to keep a well open in either unconsolidated materials or unstable rock (Fetter, 1994).

Well cuttings: A type of drilling sample in which the material is crushed and broken into particles.

Well data: A concise statement of the available data regarding a well; a full history or day-by-day account, from the day the location was surveyed to the date production ceased (AGE, 1980, p. 699).

Well development: Application of a surging or brushing process to a well for the purpose of drawing fine material from the aquifer next to the well and increasing well yield (U. S. Army Corps of Engineers, 1991).

Well efficiency: Ratio of the actual specific capacity after 24 hours of pumping to the theoretical specific capacity at that time (U. S. Army Corps of Engineers, 1991).

Well filter: See *screen.*

Well function: Mathematical function by means of which the unsteady drawdown can be computed at a given point in an aquifer at a given time due to a given constant rate of pumping from a well.

Well intake: The well screen, perforated sections of casing, bottom of casing, or other openings through which water from a water-bearing formation enters a well (Rogers et al., 1981, p. 434).

Well log: A chronological record of the soil and rock formations encountered in the operation of sinking a well, with either their thickness or the elevation of the top and bottom of each formation given. It also usually includes statements about the lithologic composition and water-bearing characteristics of each formation, static and pumping water levels, and well yield (Rogers et al., 1981, p. 434).

Well point: A hollow vertical tube, rod, or pipe terminating in a perforated pointed shoe and fitted with a fine mesh wire screen, sometimes connected with others in parallel to a drainage pump and driven into or beside an excavation, to remove underground water, to lower the water level and thereby minimize flooding during construction, or to improve stability (AGE, 1980, p. 699). Also used to obtain a small water supply from a shallow unconfined aquifer.

Well screen: A tubular device with either slots, holes, gauze, or continuous wire wrap; used at the end of a well casing to complete a well. The water enters the well through the well screen (Fetter, 1994).

Well yield: The rate at which a well will yield water either by pumping or free flow (After Rogers et al., 1981, p. 439).

Wellhead protection area: A protected surface and subsurface zone surrounding a well or well field supplying a public water system to keep contaminants from reaching the well water (U. S. Environmental Protection Agency, 1994).

Withdrawal: The act of removing water from a source for use; also, the amount removed (AGE, 1980, p. 703).

Xerophyte: A plant adapted to dry conditions; a desert plant (AGE, 1980, p. 706). A desert plant capable of existing by virtue of a shallow and extensive root system in an area of minimal water (Fetter, 1995).

Yield: (1) The quantity of water, expressed as a rate of flow, that can be collected for a given use or uses from surface or groundwater sources. The yield may vary with the use proposed, with the plan of development, and also with economic considerations. (2) Total runoff. (3) The streamflow in a given interval of time derived from a unit area of watershed, usually expressed in cubic feet per second per square mile. (4) The amount of material ultimately produced by a given process, for example, volume or weight of sludge produced, (Rogers et al., 1981, p. 440). Also the rate of pumping water from a well or well field without lowering the water level below the pump intake. See *water yield; well capacity.*

Zone of capture: The area surrounding a pumping well that encounters all areas or features that supply groundwater recharge to the well (National Academy, 1994).

Zone of influence: The area surrounding a pumping or recharging well within which the water table or potentiometric surface has been changed due to the well's pumping or recharge.

Zone of saturation: The zone in which the permeable rocks are saturated with water under hydrostatic pressure. Water in the zone of saturation will flow into a well and is called groundwater. That part of the Earth's crust beneath the regional water table in which all voids, large and small, are filled with water under pressure less than atmospheric (Langbein and Iseri, 1960). *Saturated zone.*

REFERENCES

Allaby, M., 1977, *A dictionary of the environment.* New York: Van Nostrand Reinhold Company, 532 pp.

Alley, W. H., Riley, T. E. and Frank, O. L., 1999, *Sustainability of groundwater resources.* Denver: U.S. Geological Survey, 79 pp.

American Society of Agricultural Engineers, 1967, *Glossary of soil and water terms,* Nomenclature committee, Soil and Water Division, Special Pub, 45 pp.

American Society of Civil Engineers, 1949, *Hydrology handbook, Manual of Engineering Practice*, no. 28, 184 pp.

ASTM (American Society of Testing and Material), 1994, *ASTM standards on groundwater and vados zone investigations*, 432 pp.

Bates, R. L., and Jackson, J. A., eds, 1987, *Glossary of geology,* third edition. Alexandria, VA: American Geological Institute.

Bowen, R., 1980, *Groundwater.* New York: John Wiley and Sons, Halsted Press Division, 227 pp.

Bryan, K., 1922, *Erosion and sedimentation in the Papago country, Arizona, with a sketch of the geology*, U.S. Geological Survey Bull, 730-B, 19–90 pp.

Bucksch, H. 1996, *Dictionary geotechnical engineering.* Heidelberg: Springer, 699 pp.

Cohen, R. M., and Mercer, J. W., 1993, *DNAPL site evaluation.* Boca Raton, FL: CRC Press.

Colby, B. R., Hembree, C. H., and Jochens, E. R., 1953, *Chemical quality of water and sedimentation in Moreau River drainage basin, South Dakota*, U.S. Geological Survey Circular 270, 53 pp.

Dominico, P. A., and Schwartz, F. W., 1990, *Physical and chemical hydrogeology.* New York: John Wiley and Sons, 824 pp.

Dunne, T., and Leopold, L. B., 1978, *Water in environmental planning.* San Francisco: W. H. Freeman and Company, 818 pp.

Durfor, C. N., and Becker, E., 1964, *Public water supplies of the 100 largest cities in the United States, 1962*, U.S. Geological Survey Water-Supply Paper 1812, 364 pp.

Environmental Protection Agency, 2001, *Terms of environment*, U.S. Environmental Protection Agency.

Erbe, N. A., and Flores, D. T., 1957, *Iowa drainage laws* (annot.), Iowa Highway Research Board, Bulletin 6, 70 pp.

Ferris, J. G., Knowles, D. B., Brown, R. H., and Stallman, R. W., 1962, *Theory of aquifer tests*, U.S. Geological Survey Water-Supply Paper 1536-E, 174 pp.

Fetter, C. W., 2001, *Applied hydrogeology.* New York: Macmillan College Publishing Co., 691 pp.

Flick, G. W., ed. 1980, *Environmental glossary.* Washington, D.C.: Government Institutes, 196 pp.

Freeze, A. R., and Cherry, J. A., 1979, *Groundwater.* Englewood Cliffs, NJ: Prentice Hall, 604 pp.

Gilpin, A., 1976, *Dictionary of environmental terms.* London: Routledge and Kegan Paul, 189 pp.

Glover, R. E., 1964, The pattern of fresh-water flow in a coastal aquifer, in Cooper, H. H., Jr., eds., *Seawater in coastal aquifers*, U.S. Geological Survey Water-Supply Paper 1613W, 84 pp.

Gorden, N. D., McMahon, T. A., and Finlayson, B. B., 1992, *Stream hydrology: an introduction for ecologists*, New York: John Wiley and Sons, 526 pp.

Groundwater Subcommittee (of the Federal Interagency Committee on Water Data), 1989, *Federal glossary of selected terms: subsurface-water flow and solute transport*, U.S. Geological Survey, Office of Water Data Coordination, 38 pp.

Harrison, W. E., and Testa, S. M., 2003, *Petroleum and the environment.* AGI Environment Series, 64 pp.

Holzer, T. L., 2009, *Living with unstable ground.* AGI Environment Series, 64 pp.

Horton, R. E., 1933, *The role of infiltration in the hydrologic cycle*, American Geophysical Union, 446–460 pp.

Hutchinson, D. E., compiler, 1970, Resource conservation glossary, *Journal of Soil and Water Conservation*, vol. 25, no. 1 (January–February 1980).

Lamb, T. W., and Whitman, R. V., 1969, *Soil mechanics.* John Wiley and Sons.

Langbein, W. S., and Hoyt, M. G., 1959, *Water facts for the nation's future.* New York: The Ronald Press Company, 288 pp.

Langbein, W. S., and Iseri, K. T., 1960, *General introduction and hydrologic definitions*, U.S. Geological Survey Water-Supply Paper 1541-A, 29 pp.

Langer, W. H., Drew, L. W., and Sachs, J. S., 2004, *Aggregate and the environment.* AGI Environment Series, 64 pp.

Linsley, R. K., Jr., Kohler, M. A., and Paulhus, J. L. H., 1949, *Applied hydrology.* New York: McGraw-Hill, 689 pp.

Liptak, B. G., ed., 1974, *Environmental engineers' handbook: volume 1, water pollution.* Radnor, PA: Chilton Book Company, 2,018 pp.

Lohman, S. W., and others, 1972, *Definitions of selected groundwater terms, revisions and conceptual refinements,* U.S. Geological Survey Water-Supply Paper 1988, 21 pp.

Loynachan, T. E., and others, 1999, *Sustaining our soils and society.* AGI Environment Series, 64 pp.

Maxey, G. B., 1964, Hydrostratigraphic units, *Journal of Hydrology,* vol. 2, 124–129 pp.

Meinzer, O. E., 1923, *Outline of groundwater hydrology,* U.S. Geological Survey Water-Supply Paper 494, 71 pp.

Meinzer, O. E. 1927, *Plants as indicators of groundwater,* U.S. Geological Survey Water-Supply Paper 577, 95 pp.

Monroe, W. H., 1970, *A glossary of karst terminology,* U.S Geological Survey Water-Supply Paper 1899-K, 26 pp.

National Academy, National Research Council, 1993, *Alternatives for groundwater cleanup.* Washington, D.C.: National Academy Press, 315 pp.

Paylore, P., ed., 1974, *Phreatophytes: a bibliography.* U.S. Department of Interior Office of Water Resources Research, 277 pp.

Pfannkuch, H. O., 1969, *Elsevier's dictionary of hydrogeology.* New York: Elsevier Publishing Company, 168 pp.

Robinson, T. W., 1958, *Phreatophytes,* U.S. Geological Survey Water-Supply Paper 1423.

Rogers, B. G. et al., 1981, *Glossary: water and wastewater control engineering.* American Water Works Association.

Rantz, S. E., and others, 1982, *Measurement and computation of streamflow: volume 1, measurement of stage and discharge,* U.S. Geological Survey Water-Supply Paper 2175.

Seaber, P. R., 1986, Evaluation of classification and nomenclature of hydrostratigraphic units, *EOS Transaction of the American Geophysical Union,* vol. 67, no. 16, 281 pp.

Solley, W. B., Chase, E. B., Mann, W. B., IV, 1983, *Estimated use of water in the United States,* U.S. Geological Survey Circular 1001, 56 pp.

Stanger, G., 1994, *Dictionary of hydrology and water resources.* Adelaide, South Australia: Lochan, 208 pp.

Stone, A. W., and Stone A., 1994, *Wetlands and groundwater in the United States.* American Groundwater Trust, 100 pp.

Stringfield, V. T., 1966, Hydrogeology: definition and application, *Groundwater,* vol. 4, no. 4, 2–4 pp.

Studdard, G. J., compiler, 1973, *Common environmental terms: a glossary,* U.S. Environmental Protection Agency, EPA Region IV, Atlanta, Georgia, 23 pp.

Titelbaum, O. A., 1970, *Glossary of water resource terms,* Federal Water Pollution Control Administration, 39 pp.

UNESCO, 1970, *International legend for hydrogeological maps,* International Association of Scientific Hydrology, Publication No. 98, 101 pp.

UNESCO, 1974, *International glossary of hydrology,* QM/OMM/BMO-No. 385, first edition. 393 pp.

U.S. Army Corps of Engineers, 1991, *Glossary of hydrologic engineering terms.* Davis, CA: Hydrologic Engineering Center, 135 pp.

U.S. Department of the Interior, 1986, *Glossary of selected water-resource and related terms.* Reston, VA: Water Resource Division, U.S. Geological Survey, 141 pp.

U.S. Environmental Protection Agency, 1975, *Manual of individual water supply systems,* EPA 430/9-74-007, 153 pp.

U.S. Environmental Protection Agency, 1994 and 2001, *Terms of environment, glossary, abbreviations, and acronyms,* U.S. Environmental Protection Agency EPA 175-B-94-015.

U.S. Geological Survey, 1985, *National water summary 1985*, U.S. Geological Survey Water-Supply Paper 2300.

U.S. Geological Survey, 1986, *National water summary 1986*, U.S. Geological Survey Water-Supply Paper 2325.

U.S. Geological Survey, 1989, *Federal glossary of selected terms subsurface-water flow and solute transport*, U.S. Geological Survey, 38 pp.

U.S. Geological Survey, 1993, *National water summary 1990–91*, U.S. Geological Survey Water-Supply Paper 2400.

USSCS (U.S. Soil Conservation Service), 1971, *National engineering handbook*, Section 4, Hydrology, Chapter 22, Glossary.

Vandas, S., Winter, T. C., and Battaglin, W. A., 2002, *Water and the environment.* AGI Environment Series, 64 pp.

Wang, L. K., ed., 1974, *Environmental engineering glossary: Over 3800 definitions.* Buffalo, NY: Calspan Corporation, 439 pp.

Weight, W. D., 2008, *Hydrogeology field manual.* New York: McGraw Hill, 751 pp.

Wilson, W. E., and Moore, J. E., 1998, *Glossary of hydrology.* Alexandria, VA: American Geological Institute, 248 pp.

Wisler, C. L, and Brater, E. P., 1959, *Hydrology.* New York: John Wiley and Sons, 408 pp.

World Meteorological Organization, 1974, *International glossary of hydrology.* WMO Publication.

Appendix A : The Ideal Project—Its Planning and Supervision

John E. Moore and Hugh Hudson
U.S. Geological Survey (USGS) Staff Hydrologists,
USGS Open File Report 91-224

A.1 INTRODUCTION

Project orientation, planning, supervision, and timely completion are all areas that are receiving increasing attention at regional, division, and director levels. This emphasis has been strongly indicated by the director's public statements regarding improving the usefulness of survey reports and by the publication of priority guidelines in water resources division memorandums. Valid criticism has been directed at the geological survey regarding lack of timeliness and relevance of reports. This can be corrected by improved project design, supervision, and management.

Assume that funds and personnel will be assigned to those projects that meet the criteria of technical quality, relevance to current and potential needs, and adequate management to ensure that these needs are fulfilled.

Before getting into specifics, the following is given for perspective: Within the total water resources division program, there are about 1,500 projects. The number of reports from these projects that were prepared for the director's approval in 1970 was more than 800. In 1950, there were only 300.

In the 12-state Rocky Mountain region, there are about 230 interpretive projects that will produce reports. The total funding of work in this region in fiscal year 1972 was $19 million. Interpretive projects used $10 million of the money (area appraisal and applied research). Area appraisal projects consumed nearly $4.5 million in 1972. Clearly, project planning and management represent substantial investments in people and money and the call for a businesslike approach. The purpose of this appendix is to list guidelines for better project planning and supervision. Obviously these guidelines should not be considered as a panacea to resolve all the problems related to projects. Before presenting these guidelines, we offer the following as our definition of the "ideal project."

A.2 CHARACTERISTICS OF THE IDEAL PROJECT

A.2.1 SPECIFIC OBJECTIVES

The objectives should point to the salutations of specific problems. If the objectives are not clear, the approach cannot be determined and the project is in danger of aimless roaming in search of its objective. Fuzzy definition of the project goal leads to uncertainty about the merits of each step taken during the project execution. Uncertainty frequently leads to time misspent in collecting, cataloging, and interpreting trivia.

A.2.2 LIMITED DURATION

Ideally, the length of a project should be 24 to 36 months. Longer projects frequently run into problems of completing reports on schedule and maintaining staff continuity. If this span of time appears impracticable, the project should be broken into phases of relatively brief duration with a specified goal and report for each phase.

A.2.3 FULL-TIME AND CONTINUOUS STAFFING

Full-time and continuous staffing is essential to efficient project operation and management. The probability of staffing interruptions by transfer and priority changes by the cooperator is decreased with shorter projects. Full-time participation, particularly by the project chief, is almost essential. For the project chief to be required to divide his or her time between projects is patently inefficient, and he or she may be tempted to play one deadline off against the other and delay completion of both projects. If possible, an interdisciplinary team approach is recommended. Many of the new projects in the water resources division include full- or part-time input from groundwater, surface water, and water quality disciplines. An ideal project requires a widely diversified experience, interest, and capability. The project leader is responsible for assembling, guiding, and using the technical talents of the staff.

A.2.4 ADEQUATE FUNDING

Lack of adequate funds is probably the major cause of failure. Although the absolute necessity of adequate funding cannot be denied, inadequate funding is a surprisingly common pitfall. We tend to be somewhat overzealous in "selling" projects and succumb to the temptation to make the job more attractive to the cooperator by cutting costs to the bone. Financing must be at a level adequate for achieving stated goals. Continuing surveillance of progress by the district chief is required so that the cooperator can be advised if the original goals are within budget constraints. If financing is a basic problem to successful completion, efforts should be made immediately to revise either the objective or the budget level. Symptoms of an underfunded project are frequent cost overruns, slippage of completion date, and a substandard technical report.

A.2.5 MEETING OBJECTIVES

It goes without saying that the ideal project is completed on schedule, is technically acceptable, and meets the stated objectives. The project should produce reports that reach and are understood by the intended audience. Is the ideal project attainable? Emphatically yes, with proper attention to the details that go into planning, supervision, and report management.

A.2.6 PROJECT PLANNING

There are no hard and fast rules for planning a project. Many planning details depend on the uniqueness or difficulty of the job and the experience available within the district from similar projects. If the project under consideration is a study of a county adjacent to another county that was just studied, the planning phase may be a relatively simple modification of the earlier study, provided the earlier experience was successful and documented. Documentation is a requirement for adequate management and will be discussed in Section A.3, "Project Supervision." Project planning usually begins when the project proposal is prepared for the cooperator. The district should list specific objectives, point out the hydrologic complexities in the area, and list the major water problems. The district should obtain assistance from the regional office, research projects, or other districts for review of these proposals.

A.2.7 PLANNING REPORT

Many districts prepare a preproject planning report before any fieldwork is started. Some districts prepare the report as a separate project, while others put aside the first 3 to 6 months of the project to prepare a planning report. A planning report is highly recommended for projects that have had no predecessor in the district and for those that are above average in difficulty. The basic planning report should include (at a minimum) a clear statement of the objectives, the proposed approach, a conceptual hydrologic description, available data, data needs, work schedule, report plans, and references. The report should receive a detailed review by the cooperator, regional office, and in some cases division staff members. Some districts have had success in using a brainstorming technique to prepare plans of the report. For example, a group of hydrologists with diverse interests and backgrounds is assembled. They express the possible objectives, approaches, and project priorities. The cost of a project planning report ranges from $3,000 to $8,000 and commonly requires 1 week to 3 months to complete.

The North Dakota district recently prepared such a report. Projects that had been routinely requested by the cooperator were the traditional county groundwater reconnaissance studies. Then, the Corps of Engineers came up with plans for a reservoir on the Cheyenne River overlying an important aquifer. The question the Corps asked was what effects the reservoir would have on the local and regional groundwater environment. A secondary question was what would be the effects of the proposed reservoir on nearby seeps and springs. The district office developed a planning report to prepare for this project. It included the following: introduction, purpose of study, hydrogeology, method of study, available data, data needs, estimated costs, work schedule, selected references, a map showing the location of the study area, and a hydrogeologic section. After the report was initially drafted, a meeting was held in Denver involving representatives of the Bismarck office, the regional staff, and two consultants from the Arkansas district who were chosen because of their experience with a similar problem. The original work plan was then modified on the basis of advice and recommendations obtained at this meeting. The revised work plan served as the basis for preparing (and became a part of) the formal project proposal. About 3 months elapsed between the inception of the project and its approval. The time could have been shortened considerably, if necessary.

Planning major projects in New Mexico is done in a slightly different way. The method used is a preproject. The purpose of this project is specifically for planning. The project chief is assigned the job of assessing the problem, the hydrologic situation, developing the conceptual model, reviewing the literature and the state of the art, assessing the database, determining data needs, and preparing a work plan. The end product is a highly detailed project proposal that serves as the basis of the agreement with the cooperator as to costs, approach, duration of study, and type of report. The detailed project proposal is abstracted for and becomes a part of the formal project proposal. Such an approach costs about $6,000 to $12,000 and is money well spent when the final project may cost in the $300,000 to $900,000 range. Moreover, cost overruns from inadequately planned projects may consume several times the cost of detailed planning.

A.2.8 TECHNICAL ASSISTANCE

The North Dakota district's project report was substantially changed and improved because of consulting assistance provided by the regional office and by the Arkansas district. Much of the cost of the assistance was paid from the region's consulting fund. Districts should make use of the funds to review project plans during the formulation stage. Project personnel should enlist the aid of other district personnel, the branches, and research specialists in the design of quantitative studies. Where predictive models are contemplated, the Analog Model Unit, the Hydrologic Systems Laboratory Group, or similar research group should be consulted for technical advice beginning with the project planning. Where technical expertise in the project needs bolstering, consultations or short assignments by appropriate individuals should be sought. Such needs should be identified during the project planning phase.

A.2.9 IDENTIFICATION OF SPECIFIC OBJECTIVES

The definition of specific project objectives is probably the most important part of planning. It is recommended that a list of desirable objectives be prepared, and then those objectives that are practical to achieve be selected. Finally, the objectives should be balanced with the need for information in the study area. The selection of goals should be based on an awareness of the complexity of hydrologic and water supply problems. The most critical unknowns should be tackled first. The limits of the project area, the information needed, and the type of report should be established during the first few weeks of the project.

A.2.10 DOCUMENTATION OF PROJECT

The preparation of a formal project description should be made by the project chief. There are times when the project chief is not on board or is not selected. He should be given the opportunity to review, modify, and otherwise imprint the project with his own personal touch. The preparation of project documents, as an administrative chore, remote from the project chief, is strongly discouraged. These documents should be used to prepare the work plan and budget.

A.3 PROJECT SUPERVISION

The ideal project is now underway and its plan becomes a management tool. The following is a list of general guidelines for the supervision of projects. It includes guidelines for the project chief and district supervisors.

A.3.1 WORK PLAN

A detailed work plan containing a list of the major items of project work, completion dates, manpower requirements, and expenses should be prepared by the project chief during the first part of the project (1 to 3 months). It is prepared after the needs for data, research support, and special studies have been defined. The work effort should first be subdivided into logical units with realistic completion dates for each. The work plan should include a listing of maps, tables, and other items to be generated by the project.

A.3.2 PROJECT CONTROL

The project chief should be required to give an oral or written progress report to his or her supervisor on a regular schedule. Some districts require a report on progress and plans each month. One very useful method of keeping track of progress is to prepare a project progress chart. Actual progress is charted against the listed completion date shown on the work plan. The chart serves two purposes: (1) it provides a visual display of progress, and (2) it gives an early warning of schedule slippage. Another use of the chart is that it provides a basis for estimating cost and time requirements for future projects.

A.3.3 RECONNAISSANCE PHASE

Reconnaissance work during the early phases of a project should identify variability of hydrologic systems, data availability, and the principal controls on the occurrence and movement of water. Reconnaissance information should be used to update project work plans, guide intensity, and distribute field effort to define the significant unknowns.

A.3.4   Technical Quality Control

A systematic technical review schedule is an essential element of effective project management. It is the responsibility of the supervisor to review the technical aspects of the project frequently. The review should consider the progress, plans, and resolution of objectives. If needed, the work plan should be revised and work effort and goals rescheduled. You have no doubt heard of the district chief who gave a project leader an assignment and said, "Here's your project. Now don't let me see you again for 3 years." It probably never happened, but there are indications of infrequent or irregular internal project reviews within the district. Effective management requires close contact with the project staff. This contact consists mainly of periodic and regular technical reviews. A team approach to review has merit, particularly if the problem is interdisciplinary. Review at 3- to 6-month intervals, especially during the first year, is an effective way to sense problems and progress and to utilize decision points if changes appear to be in order. Such reviews may be a part of regular staff meetings or district technical seminar sessions. These reviews not only provide technical guidance but also identify the amount of time being expended on each phase of the project. Some project chiefs frequently expend too much effort on that phase of the project where they personally have the greatest interest or expertise. The supervisor should also visit the field. There is no substitute for understanding the field problems. The district chief should seek outside assistance from the region, the branches, research, or outside the survey to assist in technical review. Outside help is especially needed on projects that present a new approach for the district.

A.3.5   Oral Presentation

The project personnel should be encouraged to present talks to cooperators, technical societies, and community groups. The advantages gained from this are many. For example, it provides good public relations for the district, should improve the report, and may result in expansion or change in project objectives.

A.3.6   Professional Environment

It is the basic responsibility of the district chief to provide a productive environment for the employee. Key points here are the opportunity for the project personnel to take an active part in project planning; to freely exercise imagination in obtaining, interpreting, and presenting results; and to communicate freely on technical problems with peers in other projects, districts, and agencies. Stated in a slightly different way, project personnel should be given an opportunity for professional growth through assigned responsibility rather than through a tightly restricted set of duties. Project personnel should be made aware of their responsibilities by frequent consultations with the supervisor; continuing review of project progress by district officials results in commendations, where warranted. Project personnel should be surrounded by an attitude that stresses getting the job done.

A.3.7 Report Management

Report management is a subject that should receive separate treatment all its own; however, reports cannot be separated or ignored in project planning or supervision. Reports can be improved by giving more attention to colleague review and making them less stereotyped, releasing them more rapidly, and preparing more reports related to the water user.

A.3.8 Report Planning

Report planning is a continuing process. Some suggestions for planning reports are as follows: A preliminary report, which outlines main hydrologic features of study area (using data available) and suggests work needed to eliminate deficiencies and analytical techniques to be applied, should be prepared during the first 10 percent of project life. A series of short internal reports covering successive phases of project work are valuable for training in report writing and can be added to the final report.

Report preparation should never be handled as a chore to be done just before the project is concluded. Work on the outline and parts of the final report should be done in steps as fieldwork reaches identifiable conclusions throughout the life of the project. The project chief should submit the first drafts of the report not later than 6 months before the end of the project.

Appendix B: Aquifer Test—An Alternative Data Interpretation

J. Joel Carrillo-Rivera and A. Cardona

B.1 OBJECTIVES OF AQUIFER TEST

Data collection for aquifer tests during the last decades, and their analysis, has received considerable attention by groundwater researchers. Aquifer test was originally called *pumping test*, but the former term seems more appropriate as the objective of the test is to identify the hydraulic characteristics of aquifer units under investigation, not of a pump. A great number of papers have been written on the subject as well as textbooks edited where purely hydraulic explanations are considered. The understanding of the flow to the extraction well seems to be the key. Additional features such as chemical and overall response of the flow system involved during an aquifer test may be used as additional tools to understand observed groundwater heads in the aquifer test. This response may provide further understanding of groundwater functioning through a conceptual model of the study site under consideration that approaches that of the prototype. Consequently, this chapter puts forward an alternative view to highlight field features and data to be gathered as part of an aquifer test procedure, as well as a method of solution easily applicable to most field hydrogeological conditions.

As a general statement, the objectives of an aquifer test are (1) to identify the performance or well efficiency, (2) to estimate extraction capability in terms of yield and drawdown with extraction time, (3) to assess aquifer properties of the part of the geological units affected by the test, (4) to assist in identifying the nature of groundwater flow to the borehole, and (5) to recognize groundwater flow control on obtained water quality. These objectives are reached by the joint interpretation of the following readings to be obtained during the aquifer test: (1) the water-level fluctuation (s) in an extraction (and often in an observation) borehole recorded continuously since the extraction (t) started, and (2) the chemical and physical response of extracted water, which is recorded continuously once extraction has started.

B.2 GENERAL CONDITIONS AND DESIGN OF AQUIFER TESTS

Aquifer testing implies that some field conditions related to the well and its surrounding hydrogeology setup are ideally met to have a basic conceptual model on how groundwater is functioning when the borehole is tested. The conceptual model should be supported by additional information, most of which is to be collected in the field; the following are required:

of the extraction well
1. well design (diameter, depth, type of casing, and its location depth)
2. space to introduce a device to measure the position of the water level
3. a reliable pumping system that controls obtained yield to an error of ±5 percent
4. means for a continuous determination of the groundwater extraction rate
5. site where extracted water is to be disposed, away from the tested aquifer unit
6. a water tap to install an isolated line-cell to collect water samples and measure field data
7. distance to possible observation wells to be used during test
8. location of extraction wells and their performance
9. well lithology as well as basement rock location (depth and type)
10. lateral low-hydraulic-conductivity boundaries
11. distinguishing of heterogeneity and isotropy of tested aquifer units

of water
1. chemistry of groundwater (and surface water) in the various geological units involved
2. readings of water levels (and chemistry) prior to the test
3. distance to lateral inflow boundaries (river, lake, canal, wetland)

of the owner of the well
1. to set the pump to start and stop at the required time
2. to be sure that the pump has sufficient fuel for the required duration of the test
3. extracted water to be diverted properly

It should be kept in mind that collection of the readings obtained in the field during aquifer test procedures should be done on a continuous basis. The readings are related to the evolution of the water level (s) with extraction time (t) in extraction and observation boreholes. Obtained s–t (drawdown-time) values are paramount in standard aquifer test data interpretation. However, in this alternative interpretation of an aquifer test, basic chemical data are incorporated to assist in the analysis, implying that an additional response to the lowering of the water table is reflected in the extracted water by altering its basic chemical composition. In other words, if groundwater flow to the extraction borehole is other than horizontal, it might include changes in the chemical characteristics of the water that might be reflected in variations in temperature, Eh, electrical conductivity (EC, or total dissolved solids), pH, and dissolved oxygen (DO) in the obtained water (Huizar-Alvarez et al., 2004).

B.3   AQUIFER AND AQUIFER TEST: WHAT IS THE CHALLENGE?

Aquifer is defined as "part of a geological formation, group of formations or an individual formation that may provide the water needed for a particular use, in an economic way, with the desired quality and in sufficient quantity" (Price, 2007).

In practical terms, it is common to find geological formations that cover large regions, that is, from hundreds to thousands of square kilometers, and are by all means largely heterogeneous (their physical properties vary from place to place) and thick (i.e., from hundreds to thousands of meters in thickness). The physical properties (hydraulic characteristics) of an aquifer matrix are important as they provide information on how fast groundwater saturating this matrix flows in x, y, z dimensions (hydraulic conductivity in K_x, K_y, K_z), how much water may be obtained from a saturated portion of the aquifer matrix (effective porosity, η) represented by the response in terms of drawdown volume on the hydraulic head to a given extraction rate (storativity). These characteristics of the aquifer matrix provide the flow controls on obtained water quality during extraction time and at a particular extraction rate; that is, a larger vertical hydraulic conductivity as compared with the horizontal may induce vertical flow, rather than horizontal, into the extraction borehole. The storativity of a unit thickness of an aquifer is termed specific storage, Ss; this factor may be computed by dividing the storativity of the aquifer by the aquifer thickness (b). (For further support see Section 1.1, "Hydrogeologic Concepts.")

There are various methods in the literature on how to obtain the basic aquifer properties, namely, hydraulic conductivity, effective porosity (or specific yield), and storativity (or coefficient of storage). Largely, they are included in three wide categories according to the way data are generated: (inverse) modeling, laboratory, and field (aquifer test) methods. The first requires a strict control of other parameters involved, which is often difficult to achieve. The second usually defines aquifer properties in hand size scale samples (e.g., a few cubic decimeters). Aquifer tests assist in defining values for K, η, and S and often other related parameters of the geologic units involved such as hydraulic conductivity of semiconfining units (K'). Obtained values through such procedures have different meanings in regard to the scale of the test. Modeling could provide average values at a regional scale, opposite of laboratory methods where only a few cubic centimeters are involved; each could fulfill specific water management needs.

Aquifer tests could provide information on the hydraulic characteristics of the tested part of an aquifer as well as on the controls of groundwater flowing toward the well, and may assist in formulating the relation amount involved. An aquifer test provides information on the boundaries of the aquifer unit, heterogeneity, groundwater flow systems (Tóth, 1999) involved, well construction, and operation response.

In other words, an aquifer test could be useful for other applications such as to obtain borehole information on groundwater production in regard to the quantity and quality of the water thus obtained; this implies that it is a useful tool to define (1) well efficiency and (2) extracted water quality control. Well efficiency is related to *well loss*, which provides information on well design and its construction, as well as on operation procedures. During field observations, well loss expresses itself as a larger drawdown than if the well were 100 percent efficient; the physical explanation is that during related activities to borehole drilling, construction, and operation, the hydraulic conductivity in the immediate vicinity of the well wall is reduced.

It is often supposed that the chemical quality of the water obtained in a well is constant with extraction time; such behavior is usually observed in aquifer units of restricted thickness (i.e., a few tens of meters) where only a flow system, usually local, prevails (Tóth, 1999) and the borehole penetrates the full aquifer thickness, so the flow to the well during extraction is purely horizontal. However, a challenge is in areas where *the aquifer* is several hundreds (or thousands) of meters in thickness and several flow systems (each with contrasting water chemistry and temperature) are encountered in the vertical direction, so that changes in obtained water quality are expected with extraction time or yield. Therefore, extracted water quality may be controlled when the flow system to the borehole producing such water quality changes is understood. The latter implies that when water quality obtained in the well changes with extraction time, the flow to the borehole has an important vertical groundwater flow component.

B.3.1   WHAT TO DO BEFORE THE AQUIFER TEST STARTS?

Performing an aquifer test requires access to a constructed borehole that is to be subject to a continuous and constant rate of water extraction. Some pertinent information is also required on the borehole and its surrounding environment. Therefore, it is basic to find specific information to develop a hydrogeological conceptual model related to the borehole site from where expected response of water levels and water chemistry could be proposed; these data are as follows:

1. Well radius and depth
2. Depth to basement rock unit
3. Aquifer thickness crossed by well
4. Partial penetration by well
5. Lithology type crossed by well
6. Geological structure
7. Expected hydrogeological conditions (confined, semiconfined, unconfined)
8. Type of aquifer porosity
9. Distance and description of outcropping rock units
10. Location of streams, canals, and water bodies
11. Difference in water quality and temperature with depth
12. Water quality and temperature history of obtained water in tested well
13. History of water-level fluctuation
14. History of rainfall, or irrigation procedures, related to the testing site
15. Distance and extraction rates (Q) of other extraction wells in the vicinity
16. Nearby wells and their history of yield, water level, and chemistry
17. Present depth to static and dynamic water levels
18. Definition of expected discharge rate (Q) to be obtained
19. Means to keep a stable discharge rate
20. Discharge site and proper disposal of obtained water
21. Tap for water sample collection (before input of chlorine, if any)
22. Installation of an isolated line-cell for Eh, pH, EC, temperature, and OD
23. Means to measure the water level in the borehole (prior to and during test)

This information will be used to design a sketch with the basic conceptual three-dimensional model of the hydrogeological setting where the testing borehole is located (items 1 to 14) that will include the location of the borehole among related strata (confining and semiconfining units) that have been crossed by the drilling of the borehole and those that remained to be crossed to reach basement rock. The conceptual three-dimensional model needs to incorporate the expected water quality and temperature to be present at different levels. Items 15 to 17 provide a reference of water-level-fluctuation data to assist in defining any tendency of the water levels during the aquifer test. Distance and obtained rates of other extraction boreholes in the vicinity will help to define any influence when their extraction is stopped or resumed; the history of their yield, water level, and chemistry could prove useful to evaluate present depth to static and dynamic water level. The test and data collection need to be planned considering items 18 to 23. A data control sheet for each aquifer test location site including GPS data as well as a sketch of the hydrogeological conceptual model (items 1 to 12) could prove of great value where readings of t, s, and Q (yield) need to be included.

B.3.2 WHAT TO DO DURING THE AQUIFER TEST

As an aquifer test is in progress, one needs to obtain a continuous control on the following information during the process:

In the extraction borehole
- Measurements of depth to the water level
- Recording of EC, temperature, Eh, pH, and DO in an isolated line-cell (§)
- Maintain flow rate about 95 percent of the set value

In nearby extraction boreholes
- Recording of the time of starting and stopping extraction
- Depth to the water level at start and finish of extraction

B.3.3 WHICH PART OF THE FLOW SYSTEM IS AFFECTED BY THE TEST?

In designing an aquifer test and in analyzing results thus obtained, a definition of the hydrogeological conceptual model achieves a very important meaning. Usually, the term *aquifer* is used regardless of the tested part of the aquifer that is currently responding to the well's particular design, extraction rate, and influence of the immediate hydrogeological environment. Often a well could be drilled and constructed through a set of lithology whose description could include a particular layer or strata, that for convenience could be called *aquifer unit*.

Often an aquifer is described with an additional word or words providing data on the type of porosity, such as granular aquifer, fractured aquifer, or double-porosity aquifer. This also gives additional information that helps in defining the type and nature of flow, from the tested aquifer unit to the borehole. As this flow is expected to prevail during the test, appropriate methods to interpret aquifer test data could be used accordingly. In this regard, a wide variety of general methods have been devised to analyze aquifer test data under particular porosity conditions for granular and fractured aquifers. Additionally, specific methods have been forged to be used when a granular aquifer includes a particular hydraulic response; here the tested aquifer could be labeled as confined aquifer, unconfined aquifer, and leaky aquifer. These conditions have been analytically solved by Theis (1935), Boulton (1954), and Hantush (1956).

Various methods are used for the interpretation of aquifer test data from fractured geological units. Several authors agree on the poor level of application of the classical analytical methods made for porous media when they are applied to fractured rock aquifer units. Inconsistencies have been found in the representation of the actual flow and transport problems. There is a wide spectrum of methodology devised to solve particular fractured feature problems; however, there is an important practical limitation in the use of the various available methods to determine K, η, and Ss in fractured media, as most of them were developed to define very particular field problems in waste disposal sites and to understand questions on petroleum engineering. Most of the equations to be solved require knowledge of special features of the fractures such as location, length, width, continuity, and size. Data are required on the distribution and porosity of the blocks, formed by fractures. Also, there is a need for a net of observation points to investigate the head potential distribution in the vertical cross-section. This implies that groundwater flow problems in fractured media lack a solution through these specific methods of analysis. On the other hand, this information is usually not available in most wells constructed for groundwater extraction. Therefore, the application of most methods should be related to a specific set of field data and scale in both time and space. However, data permitting, there are various advantages in some of the available methodology such as the possibility to compute anisotropy (K_v and K_h) and storage coefficient of matrix (S_m) of blocks and of fissures (S_f). Some methods of aquifer test analysis permit the determination of K_h; when K_v is comparatively smaller, head potential data from observation boreholes are required to calculate such anisotropy (Papadopulos, 1967; Witherspoon, 1979; Gringarten, 1982). A solution to flow in particular vertical or horizontal fractures or in double-porosity media where well loss and heterogeneity are included may be found in Gringarten (1982). A solution by Boulton and Streltsova (1978) to an aquifer test carried out in a double-porosity aquifer unit, such as a fractured sandstone or a fractured volcanic tuff, could provide an estimate of the product of hydraulic conductivity (K) times aperture width (b). So, an estimate of the former is needed to accomplish an assessment of the latter. Here an awareness of the meaning of the product Kb, which in granular material has been termed *transmissivity* (T), requires further understanding of the physical processes involved. In practical terms, it should be considered that field observations could indicate that a lithological unit might have a definite thickness (b); however, the *thickness* of the unit producing water is often effectively smaller by a simple effect of heterogeneity. Consequently, the real producing thickness of a lithological unit should be used to obtain a more realistic K value of the fractured media.

Values of hydraulic properties obtained for granular material through aquifer tests reflect that the value of T could still raise several queries as in the case of unconfined aquifer units, where K, η, and S results could vary according to the location of the point where the hydraulic head is observed (Rushton and Howard, 1982). A computation where the hydraulic conductivity is obtained for a differential sample (i.e., in a laboratory) of an aquifer could not lead to the same value of T (if obtained from aquifer test interpretation) when multiplied for the whole thickness of unit involved.

Care must be taken when solving aquifer test data from a fractured aquifer unit with standard type curves for granular material; the value of T thus obtained includes severe limitations. Should a fractured aquifer unit have individual fractures uniformly distributed along a thickness of 10 m, with an aperture of 0.1 mm, a K value of 10^{-4} ms^{-1}, and η of 0.0001, then if an aquifer test is carried out and such value of K is obtained (with a type curve method for granular material) and if it is not possible to define a porosity figure (i.e., there is a lack of s–t data collected in an observation borehole), the presumed "conservative" porosity value will be definitely larger, possibly of 0.01. Consequently, the computed flow velocity would be about 100 times lower under a similar hydraulic gradient than the actual velocity. Here, an important issue is the knowledge of the aperture of the fracture.

Similar queries have risen for the determination of S when obtained from differential samples through the diffusivity concept as applied through the consolidation coefficient C_c:

$$C_c = T/S$$

or

$$C_c = Kb/S_s b$$

As an elementary unit is considered, the value of K and C_c may be obtained from a soil mechanic test, and an S_s value may be reached accordingly. However, the larger the thickness considered, the closer this value of S attains unity. In other words, the volume of water computed could reach a value larger than the volume of the aquifer involved, which is mathematically correct but lacks a physical meaning.

The appropriate application of any available method requires the knowledge of the aquifer unit(s) currently under investigation and an understanding of groundwater flow to the tested borehole.

The depth to basement (rock unit) is useful in devising the type of flow system (local, intermediate, regional) that could be developed in the area (Tóth, 1999). Therefore, the thickness of the aquifer crossed by the borehole becomes paramount; should the borehole partially penetrate the aquifer, the larger the extraction rate and the larger of K_v as related to K_h, the more important the vertical components of flow become. The lithology crossed by the borehole provides an insight to the development of such flow; for instance, when the flow to the borehole is controlled by fracture porosity, the related structure indicates the presence of preferential paths such as along a fault plane. In contrast, an aquifer unit with granular porosity that is not affected by any structure could diminish the vertical components of flow by partial penetration (mainly when K_h is higher than K_v). The presence of an aquifer unit covered by a clayey saturated confined unit with low vertical hydraulic conductivity could provide a leaky response to a borehole when the aquifer and containing unit are hydraulically connected (i.e., the potentiometric surface remains above the base of the confining unit). This mechanism of vertical flow has been defined in the literature (Hantush, 1956) and could be more emphasized when a chemical response of such leaky inflow is identified by its chemical signature (Huizar-Alvarez et al., 2004).

If horizontal flow toward the extraction well prevails during the aquifer test, the upper limit of the tested volume of the aquifer unit is the cone of depression in the potentiometric surface (centered at the extraction borehole); the lower limit is a plane bounded by the depth of the borehole (i.e., basement rock unit). It is important to define the possible influence from boreholes in the vicinity that could alter the water-level response in the area of influence when halting or starting groundwater extraction. The former could diminish observed drawdown with time since extraction began; the latter could enhance observed drawdown. Consequently, the history of extraction regime (and drawdown) in nearby bores could prove useful to define such influence in observed time-drawdown data.

In a similar fashion, streams, canals, and water bodies that are hydraulically connected within the tested aquifer volume affect the $s-t$ results, reducing the rate of drawdown during the test; a similar response could be obtained when rainfall or excess irrigation water reaches the water table. An opposite reaction is obtained when the cone of depression encounters a rock unit of contrasting low hydraulic conductivity. Distance and description of water bodies and outcrops of unit rocks as well as history of water usage on the land site will provide useful data in the interpretation of collected time-drawdown (t-s) data.

In general, a reduction in the rate of drawdown with extraction time could be the result of four basic scenarios:

- a well penetrates the upper part of the aquifer and several groundwater flows are found vertically, mainly when warmer water is located at depth (Carrillo-Rivera, Cardona, and Moss, 1996)
- clayey strata are in the tested site that through a leaky effect function as a semiconfining layer (Hantush, 1956)
- the extraction in a well located within the influence of the tested well is halted
- a water body or stream hydraulically connected to the tested site or irrigation water reaches the water table during the aquifer test duration

When any of such scenarios is met in a particular test, observed drawdown rate will be reduced, often producing difficulties in the appropriate interpretation of the particular control due to the "recovery" of observed dynamic water level. Aquifer test results could be more adequately interpreted if measurements of field physico-chemical parameters are collected and assisted by particular water quality analyses that might be performed on collected samples (i.e., trace elements found in the geological environment of interest).

A history of water-level fluctuation and expected discharge rate could assist in acquiring related paraphernalia to carry out the test: measurement of depth to water level, definition and measurement of discharge rate, and drainage facility requirements to control extracted groundwater.

It is advisable to visit the site and the general hydrogeological environment before the test to find access to the extraction well (and observation well), to cross-check the GPS location, and to examine the set-up of the well—mainly to verify that it is feasible (1) to use an electric tape or other water-level recording device, (2) to measure the extraction yield, (3) to determine the correct disposal of the extracted water, (4) to connect an isolated line-cell (before the input of chlorine), and (5) to verify the radius and depth of a well.

B.3.4 Is Water-Level Difference Similar to Drawdown?

An important feature of aquifer test procedures is to obtain drawdown variations with time since extraction (or recovery) began. This is usually acquired by comparing the static water level with levels observed during extraction or recovery and is given in terms of depth to the water level (or steady-state dynamic level). This general operation is applicable in shallow aquifers where extracted groundwater has a constant temperature (not producing density difference effect of more than 5 percent) and salinity. This means that the flow of groundwater to the extraction zone of the borehole is produced solely by the hydraulic gradient resulting from the lowering of the water level in the extraction borehole by pumping.

When groundwater is extracted in a well that taps an aquifer with different flow systems at depth, which implies water temperature increases with depth, this condition induces warmer water to upper aquifer levels; here, hydraulic head difference determination is a required adjustment other than by subtracting water-level variations. The reason is that *drawdown* is a measure of the difference in hydraulic head, which is not the case when an aquifer unit has water in the vertical scale with contrasting density values. In a strict sense, this difference would have to be measured at the bottom of the aquifer. In other words, if natural groundwater velocity is considered to be negligible, the hydraulic head (Φ) depends on the elevation of the observation point above a reference level and the weight of the water column as defined by Hubbert (1940):

$$\Phi = zg + \int_{p_o}^{p} \frac{dp}{\rho}$$

where:
   Φ, (potential) hydraulic head
   z, elevation of observation point, in relation to reference level
   g, acceleration of gravity
   p, pressure of water
   p_o, atmospheric pressure
   ρ, fluid density

This implies that the hydraulic head (and drawdown) could be obtained if a pressure device is placed at the base of the aquifer (on the basement rock unit) to measure the pressure of the unit water column. Any change in head (drawdown) will be measured as a difference in pressure of the water column (instead of a difference in water-level elevation), which will be given by the change in water density resulting from any change in the amount of dissolved salts and water temperature difference on the overlying water column. Consequently, when the flow conditions produced by the extraction change the temperature of the water column below the extraction borehole, such a new flow condition requires the correction of observed water levels to obtain drawdown values.

Direct measurement of temperature (or salinity) with depth by means of logging could prove useful. This procedure is often difficult to carry out in the extraction borehole, and mainly to the maximum depth of the aquifer. Here, the application of geothermometers is a valuable tool to chemically estimate the minimum equilibrium groundwater temperature at depth. The use of such a value under prevailing geothermal gradient provides an adequate depth estimate of groundwater flow, which could be cross-referenced with lithology records in regard to aquifer thickness (and depth of flow).

Such correction is by all means relevant. For instance, examine when shallow water at 25°C is displaced by deep water at 75°C, a case that is common when the aquifer thickness is about 2,000 m, the difference in density of warm water to that of cold is similar to that between sea water to fresh water (0.9970 gr_cm^{-3} and 0.9765 gr_cm^{-3}, respectively). Therefore, the difference in the measured depth to the water-level needs a correction when referring to it as drawdown in the hydraulic head, by subtraction several meters of after depth. This implies that the movement of water to the borehole has important vertical flow components.

B.3.5   May Nonhorizontal Flow to a Borehole Be Identified?

Groundwater movement in the vicinity of a well site has important vertical components of flow due to heterogeneity of aquifer units; these components become more important as the flow is closer to the extraction well. The vertical components of flow may not be satisfactorily identified from horizontal hydraulic head representation, where the disposition of flow lines could be affected by lithology contrast, difference in depth of well screen location, and the significance of different conditions of flow system as those encountered in a recharge or discharge area (Tóth, 1999). These effects are waiting to be adequately represented and included in aquifer test interpretation methodology. However, as a rule of thumb, the presence and nature of vertical components of flow during an aquifer test may be estimated with the understanding of the flow regime in the tested borehole. Usually, as far as hydraulic functioning is concerned, the only relevant available data collected during an aquifer test is the variation of depth to the water level in the well with extraction time.

When drawdown data fail to respond as a confined aquifer showing a *recovery* behavior, it could mislead actual field flow conditions. As a matter of fact, the plot of the *s–t* values will achieve a shape suggesting additional water is arriving into the cone of depression. This condition could be described as semiconfined conditions where leakage from an aquitard is taking place or when vertical flow from beneath is induced into the extraction level.

To understand the flow pattern into the extraction well, it is recommended to carry out continuous field measurements of temperature, salinity (as total dissolved solids or EC), pH, Eh, and DO in the obtained water during the test (with the assistance of an insulated line-cell), as they could be useful in identifying flow components other than horizontal flow. In fact, a possible procedure to deduce the control of incoming flow is to interpret continuous temperature and EC data. An additional understanding of the flow to the borehole could be achieved by interpreting other chemical data such as Eh, pH, DO, or specific ions (Carrillo-Rivera, Cardona, and Edmonds, 2002; Huizar-Alvarez et al., 2004).

The chemical and physical response of extracted water is considered to provide information on the importance of the hydraulic characteristics of the geological media influencing the test and the nature of the flow thus developed. This information is also required to make considerations accordingly (i.e., hydraulic head corrections by density effect) from where other hydrogeologic aspects could be included (i.e., preferential flow from a geological unit or from a particular flow system) or an assessment of the relative importance of the hydraulic conductivities in x,y,z for the various lithology units involved (i.e., if vertical flow is dominant, K_v should also be; if there is a high influence of leaky nature, K' should be treated likewise). Temperature and EC response with time since extraction began may suggest horizontal flow, a leaky effect, or the response or inflow from beneath from where the related importance of relative controlling hydraulic properties could be assessed.

An application of characterizing the various water inputs to an extraction borehole implies that several additional tools are available for possible *in-field* water quality control. For instance, groundwater flow distribution in the vicinity of a well could depend on its construction, lithology crossed, flow systems affected, and their contrasting difference in water quality. Examples as shown in Carrillo-Rivera et al. (2002) indicate in case (a) a three-dimensional radial flow to a borehole where a mixture of two flow systems prevails; case (b) suggests that warm water from beneath enters the borehole passing through granular material where chemical reactions occur precipitating fluoride, and groundwater temperature is lower due to the slow movement of water in the tertiary undifferentiated granular material; case (c) indicates that cold water is obtained but with high fluoride content as it flows from the tertiary undifferentiated granular material into the fractured media; in case (d), warm water (with high fluoride) is obtained from fractured material; and in case (e), cold low-fluoride water is reaching the well.

Certainly, the identification of the various flow components into an extraction borehole may not be acknowledged with aquifer test data analysis carried out with analytical solutions, which although rather simple in their application, prevent reaching the formulation of the correct question. In this regard numerical modeling encourages the correct question to be much more properly stated so that the solution obtained could represent field conditions more accurately.

B.3.6   Is It Difficult to Apply a Numerical Solution?

Analytical methods have been devised since 1935 to determine hydraulic characteristics of confined aquifers. This system of solution by over-imposing on a type curve ($u–W(u)$) a field data curve ($s–t$ or $s–r^2/t$) was entirely in agreement with available paraphernalia (slide ruler) for due calculations. Also, from the point of view of water sources evaluation, available analytical methods provided a solution for the amount of groundwater flowing across an aquifer under consideration, often computed to the depth of the tested borehole. In practical terms T was the important factor to estimate; K usually lacked special interest as in those years (early in the 1970s) an evaluation of flow velocity to assist contaminant displacement calculations was not usually in the agenda of the hydrogeologist. Further, the limitations to apply these methods (aquifer is to be isotropic, heterogeneous, of infinite extent, and fully penetrated by the borehole; the radius of borehole is infinitely small and water enters the well as result of the hydraulic gradient produced by extracted water) were difficult to meet in reality, so corrections have been devised to meet each particular limitation (see, for example, Hantush, 1961; Papadopulos and Cooper, 1967). Additional requirements are that confined or unconfined conditions should prevail throughout the duration of the test, which should start when the observed potentiometric level has attained steady-state conditions, and of course the extraction yield is to remain constant during the test.

The availability of personal computers in the 1980s provided an efficient additional tool to handle all due calculations of numerical modeling. This methodology provides efficiency with additional hydraulic characteristics such as anisotropy (K_h and K_v) of an aquifer and could include heterogeneity. It considers aquifer boundaries (inflow or impervious) and partial penetration by a well; it also includes large-diameter wells and well losses (Rathod and Rushton, 1991). However, a difficulty encountered in the application of numerical methods is that they usually require additional data than that of yield and readings of drawdown with extraction time. A basic lithology framework penetrated by the borehole is needed, including distance and type of boundary. The depth to basement rock unit is required and also the depth of the well that is (partially) penetrating the aquifer; the diameter of the well (or hand-dug well) is required. Anisotropy and heterogeneity could be incorporated as well as well loss. The input from an aquitard could be estimated, and the conditions during the test need to change from confined to unconfined. Changes in the rate of extraction need to be incorporated.

B.4   IS THERE AN AUTOMATIC FIELD METHOD TO INTERPRET AQUIFER TEST DATA?

For more than 4 decades after the Theis solution was available, the only means of making calculations in the field was with the use of a slide rule; consequently, all available methods were devised under prevailing paraphernalia limitations. In the late 20th century, with the accessibility of the laptop computer, solutions to several groundwater engineering problems through numerical modeling became more popular. Among available tools, Rathod and Rushton (1991) developed a powerful model to represent two aquifers and two semiconfining units; the computer code is included and is ready to be used directly from a disk drive (c:).

The use of this aquifer test tool considers data of borehole design and construction, extraction yield, and the geological and hydrological framework in the vicinity of the borehole. The usage of this computer code is not a substitute for the needed experience of interpreting a single aquifer test and step drawdown test data to estimate hydraulic parameters of geologic materials that influence the response of the potentiometric level. It is recommended to remember that the recharge figure to be considered in unconfined cases, in the model, is the variable in the hydraulic system that includes a great uncertainty in its determination. This code is considered a valuable tool that incorporates flow mechanisms that influence the response of the potentiometric surface due to extraction.

After data interpretation is carried out, it is advisable to verify the sensibility of the values of the parameters thus obtained. It should be remember that the computed solutions are to represent field flow conditions to the well, where vertical flows may be included.

B.4.1   WHICH HYDRAULIC CONDITIONS MAY THE NUMERICAL MODEL SOLVE?

Usually, type curve and straight-line methods, which are thoroughly described in the hydrogeologic literature, for confined, semiconfined, and unconfined aquifers suppose that tested aquifer units have been totally penetrated by the tested boreholes (extraction and observation boreholes). It is also presumed that the former has a diameter small enough to produce any significant water storage effect in the borehole. It is accepted also as a prerequisite in aquifer test data interpretation with analytical solutions that the aquifer has an infinite aerial extent, implying that no boundary conditions (inflow or impervious) affect the duration of the test. Of course, the solution considers that the test is carried out in a homogeneous and isotropic aquifer unit and the test started when drawdown equals zero (static conditions). Extracted yield is to be kept constant throughout the duration of the test. Available analytical solutions imply that the flow produced in the aquifer by the extracted water is purely horizontal; this implies that there are no vertical components of flow, except by those when there is flow to the aquifer unit from a semiconfined layer. Often such factors are not met in actual field conditions and influence aquifer test results; consequently, specific analytical methods have been devised to solve the groundwater flow partial differential equation incorporating corrections for the various related factors. Each factor has been dealt with in the hydrogeologic literature as a correction that needs to be applied separately to the drawdown-time data in order to, eventually, arrive at the THEIS (1935) solution once each individual restriction is solved.

These aspects influencing significantly the response of groundwater obtained from an aquifer unit by an extraction borehole have not been included in a single basic analytical solution. On the contrary, often compelled data or minimum information required is extraction yield and drawdown response with time since extraction began. However, the numerical method such as that devised by Rathod and Rushton (1991) may be used to represent almost any particular behavior of the water level in an aquifer unit or an aquifer system composed of any combination of two aquifer units and two aquitards. In this case, relevant information of the extraction borehole is required as well as that on the geological media; a preliminary understanding of prevailing flow system related to the extracting borehole could be advantageous to interpret s-t data.

Rathod and Rushton (1991) solve, by numerical techniques, the partial differential equations that describe the movement of groundwater to an extraction borehole, where a large number of hydrogeological parameters may be included. To gain an additional understanding of this specific methodology, it is suggested to read Rushton and Redshaw (1979).

Field data required for the numerical model are particularly relevant and virtually at no cost. Actual field conditions are to be considered in all due calculations. The data set incorporates information on well construction (diameter and depth), well losses, extraction regime, presence of geological-hydrological barriers, saturation thickness of aquifer units (including fractured material) and aquitards, vertical and horizontal variation of hydraulic properties, change of confined to unconfined conditions, presence of semiconfined conditions, recharge, delay gravity drainage, and degree of partial penetration by the well.

Indeed, most of the prerequisites to interpret aquifer test data are solved using an analytical model (such as heterogeneity, isotropy, full penetration, and small radius). However, the option of using this numerical model solves such conditions and incorporates other field conditions, specifically those of the well and related hydrogeological environment. In this regard the numerical model by Rathod and Rushton (1991) is an appropriate tool that solves numerically, in a single model, what would otherwise require several analytical methods.

As a matter of fact, standard aquifer test procedures usually observe the hydraulic response of the water table in a borehole. However, it has been suggested that this answer is to be supported by simultaneous field observations of the chemistry and temperature of extracted water with those of drawdown with time since extraction began. These additional observations are suggested to develop a hydrogeological conceptual model where the flow systems and geological framework are paramount in defining the hydraulic (and chemical) response observed in the extraction borehole. These means of aquifer test data interpretation could be relevant in analyzing interbasin flow and the groundwater balance in a given surface basin (Carrillo-Rivera, 2000). Numerical modeling of an aquifer test with simultaneous chemical data collection could assist in the insight of some groundwater contamination problems such as determining the nature of the flow to the extraction borehole and induced flow from a different depth to that of the particular production level.

B.4.2 Some Key Cases of Vertical Flow Component

There are a great number of hydrogeological conditions that may affect groundwater flow to a borehole during extraction; vertical flow in the immediate vicinity of the borehole is an important one. An extraction borehole fully penetrates the aquifer system to the basement rock unit. An outer boundary acting as an effective impervious barrier is present.

B.4.3 May the Numerical Model be Applied to a Fractured Aquifer?

Often the term *transmissivity* (hydraulic conductivity times saturated thickness) has been taken as a reference to define the productivity of fractured media; however, the thickness of the aperture size is usually a query. An approximation to get an estimate of hydraulic conductivity of fractured media may be obtained with a calculation of the productive aquifer thickness. The model considers the thickness of an aquifer unit; if, for example, a borehole penetrates 100 m of a fractured rock unit and the actual productive section is found between 50 and 65 m, this could be considered as an initial estimate of the productive thickness of the aquifer unit.

B.4.4 REAL AQUIFER TEST CASES

The numerical conceptual model considers radial movement to a borehole; as the model is formulated on different variables, a strict control on lithology and borehole data is a must in order to reduce uncertainties on obtained drawdown-time data from the model to represent field data. Data control is to avoid possible different flow conditions, under diverse lithology constraints that may produce a similar drawdown result with a given extraction rate.

The model has some disadvantages, as in certain situations when the heterogeneous nature of the aquifer units, or the occurrence of lateral flows within the aquifer, means that the radial flow approximation is not applicable. Another limitation is that when the upper aquifer unit is under water-table conditions, it is difficult to distinguish between an unconfined response at the water table and the confined response below the water table. Grout (1988) has developed a numerical model that includes this effect. Regardless of such limitations, the model as devised by Rathod and Rushton (1991) is considered to more appropriately represent most field conditions and to provide more adequate results than those obtained through analytical methods. The procedure proposed here intends to provide evidence on the nature of the flow to the extraction borehole that explicitly includes the lithology framework and borehole design as well as data traditionally used in the solution for analytical methods.

The response of extracted water in terms of total dissolved solids (electrical conductivity) and temperature (as well as chemistry) is additional information to define the nature of the flow toward a well. This response is considered to be adequate in defining the nature of the vertical components of flow to the well. For further details see Carillo-Rivera and Cardona (2000).

ADDITIONAL RESOURCES

Boulton, N. S., 1954, Unsteady radial flow to a pumped well allowing for delayed yield from storage, *Proc Gen Assembly, Int. As Sci Hydrol*, Pub 37, 472–477 pp.

Boulton, N. S., and Streltsova, T. D. S., 1978, Unsteady flow to a pumped well in an unconfined fissured aquifer, *Journal of Hydrology*, vol. 37, 349–363 pp.

Carrillo-Rivera, J. J., 2000, Application of the groundwater-balance equation to indicate interbasin and vertical flow in two semi-arid drainage basins, México, *Hydrogeology Journal*, vol. 8, no. 5, 503–520 pp.

Carrillo-Rivera, J. J., Cardona, A., and Edmunds, W. M., 2002, Use of extraction regime and knowledge of hydrogeological conditions to control high fluoride concentration in abstracted groundwater: basin of San Luis Potosi, Mexico, *Journal of Hydrology*, vol. 261, 24–47 pp.

Carrillo-Rivera, J. J., Cardona, A., and Moss, D., 1996, Importance of the vertical component of groundwater flow: a hydrogeochemical approach in the valley of San Luis Potosi, Mexico, *Journal of Hydrology*, vol. 185, 23–44 pp.

Gringarten, A. C., 1982, *Flow-test-evaluation of fractured reservoirs,* Geol Soc of Am Special Paper 189, 237–263 pp.

Grout, M. W., 1988, *Numerical model for interpreting complex pumping test responses*, University of Birmingham, UK, PhD thesis, May 1988.

Hantush, M. S., 1956, Analysis of data from pumping test in leaky aquifers, *Trans Am Geophys Union*, vol. 37, no. 6, 702–714 pp.

Hantush, M. S., 1961, Drawdown around a partially penetrating well, *J Hydraul Div, Proc Amer Soc Civil Engrs,* vol. 87 (HY4), 83–98 pp.

Hubbert, K. M., 1940, The theory of groundwater motion, *Journal of Geology*, vol. 48, no. 8, 85–944 pp.

Huizar-Alvarez, R., Carrillo-Rivera, J. J., Angeles-Serrano, G., Hergt, T., and Cardona, A., 2004, Chemical response to groundwater extraction southeast of México City, *Hydrogeology Journal*, vol. 12, 436–450 pp.

Papadopulos, I. S., 1967, Nonsteady flow to a well in an infinite anisotropic aquifer, In: Procc. *Dubrovnik Symposium on Hydrology of Fractured Rocks, Int As of Sc Hydrol*, vol. 1, no. 232, 21–31 pp.

Papadopulos, I. S., and Cooper, H. H., Jr., 1967, Drawdown in a well of large diameter, *W Reso Res*, vol. 3, 241–244 pp.

Price, M., 2007, *Aqua Subterránea,* Limusa Noriga Editors, 330 pp.

Rathod, K. S., and Rushton, K. R., 1991. Interpretation of pumping from two-zone layered aquifers using a numerical model, *Ground Water*, vol. 29, no. 4, 499–509 pp.

Rushton, K. R., and Howard, K. W. F., 1982, The unreliability of open observation boreholes in unconfined aquifer pumping tests, *Ground Water*, vol. 20, no. 5, 546–550 pp.

Rushton, K. R., and Redshaw, S. C., 1979, *Seepage and groundwater flow*, Chichester: Wiley, 332 pp.

Theis, C. V., 1935. Relation between lowering of the piezometric surface and the rate and duration of discharge of a well using groundwater storage, *Trans Am Geoph U*, vol. 2, 519–524 pp.

Tóth, J., 1999. Groundwater as a geologic agent: an overview of the causes, processes, and manifestations, *Hydrogeology Journal*, vol. 7, no. 1, 1–14 pp.

Witherspoon, P. A., 1979, Observation of a potential size effect in experimental determinations of hydraulic properties of fractures, *Water Res Res*, vol. 15, no. 5, 1142–1146 pp.

Further Reading

Alley, W. M., Riley, T. E., and Franke, O. L., 1999, *Sustainability of ground water resources*, U.S. Geological Survey Circular 1186, 79 pp.

Brassington, R., 1988, *Field hydrogeology*. New York: John Wiley.

Clark, L., 1988, *The field guide to water wells and boreholes*. New York: John Wiley.

Fetter, C. W., 1994, *Applied hydrogeology*. Prentice Hall, 598 pp.

Moore, J. E., and others, 1985, *Groundwater: A primer*. American Geological Institute.

Sanders, L. L., 1998, *A manual of field hydrogeology*. Prentice Hall.

Weight, W. D. 2008, *Hydrogeology field manual*. McGraw Hill.

Weight, W. D., and Sonderegger, J. L. 2000, *Manual of applied field hydrogeology*. McGraw Hill, 751 pp.

Wilson, W. E., and Moore, J. E., 1998, *Hydrologic glossary*. American Geological Institute, 245 pp.

Winter, T. C., 1999, *Ground water and surface water: A single resource*, U.S. Geological Survey Circular 1139, 76 pp.

Index

Milton Keynes UK
Ingram Content Group UK Ltd.
UKHW031149141024
449569UK00024B/932